Causation in Science and the Methods
of Scientific Discovery

Causation in Science and the Methods of Scientific Discovery

Rani Lill Anjum and Stephen Mumford

OXFORD
UNIVERSITY PRESS

Great Clarendon Street, Oxford, OX2 6DP,
United Kingdom

Oxford University Press is a department of the University of Oxford.
It furthers the University's objective of excellence in research, scholarship,
and education by publishing worldwide. Oxford is a registered trade mark of
Oxford University Press in the UK and in certain other countries

First Edition published in 2018
Impression: 5

Published in the United States of America by Oxford University Press
198 Madison Avenue, New York, NY 10016, United States of America

British Library Cataloguing in Publication Data
Data available

Library of Congress Control Number: 2018939482

ISBN 978–0–19–873366–9

Printed and bound by
CPI Group (UK) Ltd, Croydon, CR0 4YY

Preface

There is no doubt that there is causation in science. Some of the chief goals of science are understanding, explanation, prediction, and technical application. Only if the world has some significant degree of constancy, in what follows from what, can these scientific activities be conducted with any purpose. This regularity in nature is the basis of prediction, for example, which would otherwise be entirely unreliable. But what is the source of such constancy and how does it operate? How is it ensured that one sort of event tends to follow another? This is a question that goes beyond science itself—beyond the empirical data—and inevitably requires a philosophical approach: a metaphysical one. We argue in this book that causation is the main foundation upon which science is based. It is causal connections that ground regularity. They are the reason that one kind of thing typically follows another with enough reliability to make prediction and explanation worthwhile pursuits. It is causation that grounds the possibility of knowledge.

Should scientists concern themselves with what philosophers have to say? On this issue we argue for a resounding affirmative. Philosophy tells us about the nature of causation itself and thus what we should be looking for when we investigate the world. All scientists make presuppositions. They rest their endeavours upon assumptions that are not themselves empirically demonstrable through data. These assumptions are metascientific. Scientists conduct experiments, for instance: trials and interventions that change one thing in order to see what changes with it. Without a belief in causal connections, or in something being causally responsible for those changes, such experimentation would have no point because the intervention would not be productive of the result; also there would be no possible future application. Furthermore, the methods adopted or preferred by practising scientists must reveal something of what they take causation to be. At the very least, they must be looking for something that they accept as evidence of such causal connections.

For some time, philosophers have paid an interest in science, attempting to make their philosophical theories better informed empirically. Too often neglected is the opposite goal: of making scientific practice better informed philosophically. In the case of causation this is absolutely crucial for it can make a significant difference to how science is conducted. If one is persuaded that causes are difference-makers, for instance, this is what one will seek to find. If one thinks causes are regularities, one will instead look for them.

Now why should any of this matter to anything? Actually, it matters a lot. For example, one of the greatest challenges facing humanity is climate change. For all we know, this is a threat to our very existence, or at least that of a significant number of the world's population. It is hard to think of a greater problem than that. Rightly, we look to science to help us sort out the facts, many of which remain controversial. There are a number of questions we can pose, all of which should have a scientific answer. The first is simply whether climate change is real, and almost all impartial scientists now agree that it is. Other questions have then become more pressing. Did human activity produce this climate change? What were the mechanisms? And, looking forward, what are likely to be the medium- and longer-term effects of such change? Will the polar ice caps all melt? Will sea levels rise? Will the Gulf Stream switch off? Will there be crop failures? Will we starve? Can it be stopped? Can it be mitigated? All of these are causal questions. They concern what-will-cause-what and we want scientific ways of uncovering these causal facts. One of our biggest ever problems has causation at its heart, therefore: causal matters that we look to science to explain and, hopefully, control.

The aim of this book is not to offer an overview of the philosophy of causation. Other books already do this. The recent *Causality* by Illari and Russo (2014) does an excellent job of presenting an impartial survey of the available theories of causation and how they relate to science. This has freed us up for a more opinionated investigation of the topic. We think that there are certain dominant philosophical views of the nature of causation that continue to inform and influence methods, or at least influence what we think the norms of scientific discovery should be. As we hope to show, however, some of those perceived norms create a conflict when they confront the scientist in everyday practice. Scientists know that they cannot perfectly preserve a norm of objectivity, that they cannot find exceptionless regularities that are indicative of causal connections, that they cannot attain absolute certainty, that causal knowledge is defeasible, and so on. Scientific practice thus provides a hard test for our received norms and philosophical theories of causation and, we argue, gives us good grounds to overthrow and replace some of those norms. We strive, therefore, for a scientifically informed philosophy as well as a philosophically informed science.

With twenty-eight chapters, it would certainly be possible for a busy reader to dip in and out of the book. The chapters can be read in any order and in isolation. But a thesis is developed in the whole that would reward a sequential reading. If you like, you could proceed straight to the conclusion. There, we list nine new proposed norms of science that are supported by the arguments in the book. Under each norm, we list which chapters are relevant: those in which support for adoption of a new norm is justified. You could then look at the relevant chapters. Another way to read the book is to skim the keywords that we have given for each chapter, instead of an analytic table of contents, and then consult the topics that interest you the most. Although we have made it easy to read the book in these ways, we still hope that some readers will choose to follow the whole text from start to finish.

Examples are used to illustrate the philosophical points. They are drawn from a range of sciences and are kept as simple as they can be. For us, it is important that readers with different backgrounds can understand and relate the discussions to their own discipline and perhaps consider more detailed and relevant examples themselves.

The book consists of eight parts. Part I contains three chapters about the philosophical motivation for the book. Chapter 1, *Metascience and Better Science*, urges that philosophers and scientists need to address the issue of causation together, aiming towards a reflective equilibrium satisfactory both theoretically and empirically. Since the scientific methods for discovering causes reveal ontological commitments as to the nature of causation, these commitments must be made explicit and subject to critical reflection. In Chapter 2, *Do We Need Causation in Science?*, we argue that causation occupies a foundational place, underpinning the purposes of science to explain, understand, predict, and intervene. Chapter 3, *Evidence of Causation Is Not Causation*, draws a distinction between how we learn about causation (epistemology) and what we take causation to be (ontology). The rest of the book is an attempt to show that these two, although they must not be confused, are tightly and inevitably linked. How we seek to establish causation must be informed by what we think causation is.

The next four chapters, in Part II, discuss an orthodox view that causation is conceptually and epistemologically linked to perfect correlations. Chapter 4, *What's in a Correlation?*, concerns how we separate causal from accidental correlations, while neo-Humeanism struggles to make ontological sense of such a distinction. Chapter 5, *Same Cause, Same Effect*, questions the theoretical assumption that causation should be robust across all contexts, since this is not supported empirically. Chapter 6, *Under Ideal Conditions*, takes this discussion further, showing how our theoretical expectation of causal necessitation is philosophically salvaged by stipulating a set of ideal conditions. In Chapter 7, *One Effect, One Cause?*, we warn against oversimplifying causes. Typically there are multiple causes of an effect and treating them in isolation can let us miss the importance of their interaction.

Part III consists of three chapters that present an alternative approach to the Humean orthodoxy. Chapter 8, *Have Your Cause and Beat It*, introduces the notion of additive interference, where an effect is counteracted by the adding of something. This explains why causation is essentially sensitive to context. In Chapter 9, *From Regularities to Tendencies*, a case is made for understanding causes as tendencies rather than constant conjunctions, conceptually detaching the notion of causation from perfect regularity. In Chapter 10, *The Modality of Causation*, we argue that causation involves a primitive and dispositional modality, weaker than necessity but stronger than pure contingency.

Part IV promotes causal theories and mechanisms as an alternative to finding causation in regularity and repetition. In Chapter 11, *Is the Business of Science to Construct Theories?*, we argue that a causal theory is needed in addition to the data if we want something more than merely mapping the facts. Chapter 12, *Are More Data Better?*, makes a case for causal singularism, where causation happens in the concrete particular, over Humeanism, where the single instance is derived from a universal claim. In Chapter 13, *The Explanatory Power of Mechanisms*, we point to how we need qualitative and mechanistic knowledge for deep causal understanding and explanation. We here present our distinctive account of mechanisms. However, in Chapter 14, *Digging Deeper to Find the Real Causes?*, we scrutinize the reductive project of finding causal mechanisms always at lower levels of nature. We conclude that mechanisms can be higher level, and we give accounts of holism and emergence.

Part V concerns how causes are linked to effects. In Chapter 15, *Making a Difference*, we discuss the counterfactual theory and argue that it fails to account for some instances of causation. This shows that causation is not the same as difference-making. Chapter 16, *Making Nothing Happen*, develops the point that there are cases of causation where no change or event follows, but that these are some of the most important causal situations. We then look at the matter of causal chains in Chapter 17, *It All Started with a Big Bang*, and consider whether their existence shows that causation is transitive. The part ends with Chapter 18, *Does Science Need Laws of Nature?*, where we question the ontological need for universal laws in addition to intrinsic propensities and their causal interactions.

Part VI is about probability. Chapter 19, *Uncertainty, Certainty, and Beyond*, focuses on degrees of belief, such as doubt and certainty. Notably, we make room for a class of cases in which you can still be certain of something even with less evidence than before. Chapter 20, *What Probabilistic Causation Should Be*, is where we offer our theory of worldly chance, based on a distinctive account of propensities. In Chapter 21, *Calculating Conditional Probability?*, we show that our intuitive notion of conditional probability must be separated from the standard ratio analysis of $P(A|B)$, for epistemological and ontological reasons.

Part VII is primarily on the problem of external validity and contains two chapters: Chapter 22, *Risky Predictions*, and Chapter 23, *What RCTs Do Not Show*. The first of these argues that our causal predictions are essentially fallible but useful nevertheless. Indeed, any theory of prediction that did not account for its fallibility would be flawed because of that. The second points to some significant shortcomings of these large-scale population studies in dealing with causal factors such as individual variation, heterogeneity, complexity, and marginal groups.

Part VIII contains five chapters on causal discovery. Chapter 24, *Getting Involved*, develops the manipulationist insight, but mainly to argue that our causal knowledge happens in close interaction with the world, not by distanced observation. Chapter 25, *Uncovering Causal Powers*, presents our account of technological innovation, which we argue rests mainly in teasing out the hidden powers of things. Chapter 26, *Learning from Causal Failure* is a consideration of the opportunities of new knowledge that arise from unsuccessful experiments and discrepancies. Given that there is a diminishing return in confirming evidence, after a point, big breakthroughs are more likely to follow from negative results. In Chapter 27, *Plural Methods, One Causation*, we argue for a combination of epistemic pluralism and an ontological monism. The failure of analysis has led others to assume that causation in reality must be many different things. But this overlooks the other possibility: causation is one thing but primitive. We must then investigate it through its symptoms and our methods must detect those. In Chapter 28, *Getting Real about the Ideals of Science*, we challenge the motivation for idealization and abstraction in science when dealing practically with a more messy reality. There is a crisis in reproducibility, which shows how some of our expectations of science are unrealistic and based on a mistaken notion of causation.

What emerges from all these considerations is that, if we understand causation right, we should approach it in a certain way through our scientific methods. We finish by offering nine new norms for causal science, drawing together the many themes of the book.

This book is the culmination of the research carried out on the *Causation in Science* (CauSci) project at the Norwegian University of Life Sciences (NMBU) from 2011 to 2015, funded by the FRIPRO scheme of the Research Council of Norway. We thank the CauSci team (original and substitute), as well as the many project collaborators, for engaging with our work. The material was further developed through a teaching course at NMBU and we are grateful for the discussions and input from the students of PHI302 and PHI403. We have also received invaluable contributions from the CauseHealth team and collaborators. In particular, we want to thank Fredrik Andersen for raising and discussing with us many of the issues of the book. We are grateful to Johan Arnt Myrstad, the *sine qua non* of Chapter 21. A special thank you to Elena Rocca, our practising scientist in residence, for suggesting and researching many of the examples we use.

Contents

List of Figures

PART I

Science and Philosophy

1

Metascience and Better Science

1.1 What Science Is and Should Be

Science is many things. We might say that it is an activity aimed at discovering a class of important truths about the world; and then showing what can be done with that knowledge. We can think of science as a tool, a method, or, more generally, a philosophy. Central to science, perhaps even constitutive of it, is a set of norms for the correct, systematic acquisition of empirical knowledge. By a norm, we mean a standard of what ought, or ought not, to be done. For scientific norms, the concern is what we ought to do or ought not to do in order to best gather and utilize knowledge about the world. Empirical knowledge is knowledge derived from experience, as opposed to mathematical, logical, and conceptual truths, which can be discovered through use of reason.

For example, it is arguable that for knowledge to count as scientific, it needs to be objective. It cannot be just one person's view, for instance. Hence, if you claim to have found something remarkable under your bed that no one else is allowed to see, the report, in that state, will not count as scientific and will thus not be an admissible datum for any scientific theory. Being just one person's experience, this report will be classed as subjective and unverified. However, if knowledge is acquired according to the norms of science, including the norm that it be objective, then it may qualify for the status of scientific. The norms are thus important to many forms of enquiry because it often matters that a claim is scientific. It means it should be trustworthy, for instance. A justification of science could be that it is knowledge acquired through these norms, which have been accepted as the right way to know the world. Science is valuable, it might be said, because it is objective, among other things.

There are a number of norms for correct empirical discovery. It needs to be admitted, however, that these norms continue to be contested. The history of science shows multiple, ongoing debates about what is the correct scientific method. Should science start with the recording of data, for instance, or is it ever permissible—indeed inevitable—to begin with a hypothesis that you can then subject to testing? There is still worthwhile discussion to be had on this question. It may now sound surprising to hear some of the older debates, especially on subjects we think are clear-cut. Galileo (1632) offered an experimental method for science, for example; but experimentation was controversial at the time (Gower 1997: 23). We now think it obvious that science should conduct experiments. In a tradition dating back to Plato (see the allegory of the cave, *Republic* 514a–520a), however, the senses were regarded as unreliable so we were urged to trust reasoning instead, which was precise and conclusive. Even in Galileo's time, evidence from the largely unaided senses was imprecise and uncertain, so thought to have only limited use for science. Galileo compromised. Most of the experiments he proposed were thought experiments. He reasoned through what he thought had to happen, including, it seems, in the famous experiment in which two objects of the same size but different weights were thrown from a tall building. It was the use of reason that led Galileo to conclude that they would fall at the same rate. The experiment could in principle be performed 'for real' but there seemed no cause to do so, thought Galileo, since any result other than his own would be contrary to reason and therefore could not happen (Gower 1997: 30ff).

We might now think that we possess accurate measuring devices within experimental set-ups and that we can trust our observations, which are as precise as pure reasoning. But this is still contestable. Sometimes there is such confidence in the reasoning within a theory that it hardly needs any empirical confirmation. Or, as in theoretical physics, we are not yet able to empirically confirm the theory and thus have to rest the argument on purely rational considerations. Opposing this view is the thought that reason itself can no longer be trusted, in that the world needn't work in a particularly rational or intuitive way. Some results in contemporary physics seem to offend common sense, for instance. In one well-known instance, we are asked to accept that Schrödinger's cat is both dead and alive at the same time and that this explains how some aspects of the world really work. We might have to revise our view of what is logical.

We said that the norms of science were many. It may help if we offer up some initial candidate norms. All of them have an initial plausibility and they might collectively be taken as a reasonable default or starting position. We could say, then, that the following list captures much of what science ought to do:

- Be objective. For example, results should be reproducible so that others can observe the same experiment and record the same outcome.
- A theory should 'preserve the data', i.e. be consistent with it eventually, more or less.
- The more empirical evidence there is in favour of a theory, the more likely it is to be true and thus the more inclined one should be to accept it.
- Favour theories that have greater explanatory power or scope. If a new theory can explain everything that the old theory explained, and more, that is a reason to favour the new theory over the old one.
- Predictive success counts in favour of a theory in that one sign of a good or true theory is that it gets confirmed by subsequent results.

A further reason for calling these norms 'default' is that we want to leave open the question of whether they are strict, 'hard and fast', or whether they should be treated as heuristic 'rules of thumb', admitting exceptions. After all, perhaps there are some cases where you should favour a theory with narrower scope over one with broad scope, such as if the broader theory rests on implausible assumptions. We should allow that the default norms may be revised or given more precision. Predictive success is not a straightforward matter, for instance. There are problems such as confirmation bias and theory-dependence of observation where it could look like a theory has achieved predictive success but hasn't really. One would have to rule out such cases in order to preserve the integrity of the norm. If one can't do that, one might have to abandon the norm.

1.2 Why Philosophy?

This account shows that there are at least two jobs for the philosophy of science. First, we have to decide what the correct norms of science are and justify them. Is it right to favour the theory with the most empirical evidence, for instance; and if so, why? Second, once all the norms are in place, we must also consider what the best ways are of satisfying them. If, for example, we agree that science should be objective, one then has to consider whether this demands reproducibility of results (see Chapter 28). Does it mean that anyone, anywhere, should be able to follow the method and get the same outcome as the one reported? Are there other ways of satisfying the requirement of objectivity? Are any of them more important than the norm of reproducibility?

Why, it might be wondered, do we say that these questions are for philosophy to answer, rather than science itself? Can't science work it out unaided? Isn't it presumptuous to claim that science needs philosophy? There is indeed a view that philosophy is of little or no relevance to this discussion, nor is anything else outside of science. Scientism, sometimes called naturalism, is the view that every meaningful question can be answered scientifically. Insofar as philosophy is attempting a task which is non-empirical, hence non-scientific, scientism questions philosophy's right to pronounce about the nature of the world. Some proponents of this view are within science. Stephen Hawking (2011: 5) has declared that philosophy is dead because physics has answered all its substantial questions. But some defences of scientism also come from within philosophy, such as from Ladyman et al. (2007), who are following a naturalist tradition promoted by Quine (1995, for one instance). We wish to overcome any such conflict between philosophy and science. Being pro-philosophy does not mean one is anti-science. Similarly, being pro-science should not mean being anti-philosophy. Both are needed, as we aim now to explain.

First, if our task is to understand what the correct norms of science are, the answer clearly cannot come from within science itself, for any such answer would be question begging. One might be tempted by the opposite view. When asked how science should be done, someone could say that one should look at science and see how it is done, or that one should consult a trained scientist who knows exactly. But this is no good. We want to know how science ought to be conducted, which isn't the same as how it is actually conducted. If we take the norms to be definitive or constitutive of science, then clearly one can only consider the validity of those norms by stepping outside of the scientific practice itself. Consideration of the normative aspect of science is thus inherently a philosophical and abstract enterprise rather than an empirical one. The key issue is whether the current norms of a practice could be wrong or incomplete. One cannot make this judgement from within that practice.

Second, it has to be acknowledged that there is philosophy in science whether one likes it or not. Science rests on philosophical assumptions, including metaphysical ones. These assumptions cannot be proven by science itself, but only assumed, and this shows us that scientism is untenable. This is effectively admitted during an attempted defence of scientism by Ladyman et al.: 'With respect to Lowe's ... claim [Lowe (2002: 6), that "Naturalism depends upon metaphysical assumptions"], it is enough to point out that even if naturalism depends on metaphysical assumptions, the naturalist can argue that the metaphysical assumptions in question are vindicated by the success of science' (Ladyman et al. 2007: 7). This concession is decisive. We can see a number of significant points in it that are worth detailing.

First, the concession accepts the possibility that science doesn't tell us everything. It can rest upon non-scientific assumptions. It is clear that these assumptions cannot be evaluated by science itself because they must be assumed in order to have science in the first place.

Next, the concession admits the basis of a strong argument in favour of a certain metaphysical assumption or set of assumptions. It tells us that such assumptions would be justified when science rests upon them and science is a success. We will see in Chapter 2 that this gives us a good argument for the reality of causation. Science assumes the validity of both observation and intervention and there could be neither without causation. We have good grounds, then, for saying that causation is real because science rests upon it and is successful. So this is an effective argument in favour of a foundational metaphysics, grounding a successful pursuit such as science.

Another significant point is that the Ladyman et al. concession assumes a type of justification for science that cannot come from science itself but, as argued above, is a norm that can only be evaluated from outside. This is the claim that success—for instance, explanatory, predictive, and technological—is what vindicates science. Science itself cannot show that you ought to follow the practice that produces success of this kind. It depends on what you want. Different activities are successful on different grounds: grounds that have to be chosen through rational enquiry. We are entitled to ask specifically what the markers of success are in science. What exactly is success in this case?

Finally, it is worth considering how this norm of science would be assessed. Suppose the naturalist were to claim, without resorting to philosophy, that it's simply obvious that predictive success is the yardstick against which we should measure science. That may be so. Isn't it just intuitive that we should justify the worth of science, and therefore its metaphysical basis, on the grounds of its success? What more than explanatory, predictive, and technological success could one sensibly want? However, this is a defence that Ladyman et al. have already ruled out. They argue that 'as naturalists we are not concerned with preserving intuitions at all' (Ladyman et al. 2007: 12) and they go on to point out how many things in science are true but counterintuitive. One should not accept the argument (that science is validated by its success) on intuitive grounds, therefore; at least not according to defenders of naturalism.

Yet, this type of argument would not be quite so bad otherwise. Ladyman et al. cannot accept it because it undermines their defence of scientism. For someone who is not seeking to defend scientism, it could however be accepted. Indeed, so compelling is the intuition, that the authors use it very quickly as a basis for accepting another norm of science. To justify that science is seeking a 'relatively unified picture' (p. 27), they argue that the opposite would be a 'mystery'. That sounds fair enough, except for the fact that the authors had just claimed that being intuitive is no virtue in science. What, then, can they maintain is bad about something being a mystery?

1.3 Philosophical Assumptions in Science

Philosophical assumptions are inevitable in science. There are then two types of scientists: those who are aware of science's philosophical underpinnings and those who are not. Only if we are aware of what we have assumed are we able to reflect critically upon its various aspects and ask whether it really gives us a sound foundation. Without such awareness, our scientific conclusions might contain erroneous presuppositions that do not serve us well or withstand rational scrutiny.

Let us look at one way in which this might happen and, in doing so, move towards the central topic that occupies us in this book: the scientific discovery of causation. We should also understand some of the differences between philosophical and scientific approaches to knowledge of causation. While science is empirical and deals with the concrete, such as particular matters of fact, recorded as data, philosophy deals with the general and abstract. Thus, in science, including social science, we might want to find out what causes what: whether a particular genotype makes a person susceptible to cancer; whether heat increases or decreases the tensile strength of a certain metal, or whether a reading programme improves the rate of social mobility. These are all particular causal hypotheses, even though none of them uses the term 'cause' explicitly. They could easily have done so ('does reading cause an increase in social mobility', and so on). Philosophy—in particular metaphysics, which studies what there is in the most general terms (Mumford 2012)—will look at what is common to these three scientific instances. They all claim that some A is causing some B. This shows that philosophy is abstract in the sense that its concern is with the nature of causation, irrespective of what the causes and effects are. Philosophy thus studies what it is for one thing to cause another, abstracted away from the particular things that are causing and being caused.

A naturalist might claim that we can just let science tell us what it is for one thing to cause another, but that would be a mistake. Science cannot do this without ceasing to be science. To do so is to go beyond the data and enter into philosophy. Science can propose theories of what causes what, based on the data, but it exceeds the norms of science to comment on the largely non-empirical matter of what causation is itself.

Naturalists find themselves in a trap, then. One can only say that A causes B if one has an account of what it is for one thing to cause another; but science unaided by philosophy cannot judge on that matter. For this reason, some scientists decline ever to make causal claims. Causal claims are rare within a certain tradition in epidemiology, for instance, even though such a science would seem pointless unless it is discovering causes (Broadbent 2013). It might be thought that the scientific task is complete once the data are recorded: perhaps a raised incidence of some disease upon increase of another variable. A naturalist reaction to the trap is, then, to proclaim that there is no causation in science, or none in one particular science. This question is the topic of Chapter 2. We can say here, however, that this sort of claim must be based on an assumption of what causation is and is not (Kerry et al. 2012, Andersen et al. 2018). Based on that assumed view of causation, one may just conclude against its presence. But what if the nature of causation is not as the naturalist assumed? Perhaps no regularities or determinacies are found in physics, but that only counts against the presence of causation if causation has to be deterministic and issue in regularities. The problem here is that we can only find the right or best account of the nature of causation and all its features by doing at least some philosophy. The naturalist is trapped by a refusal to do any substantial metaphysics, which hinders progress.

There are of course many other notions in science that benefit from a higher-level understanding. What is an explanation, for instance, a law of nature, or a natural kind? Only with a mature understanding, of what these things are supposed to be, can one accept or reject them in science. One could not say, for instance, that there are no natural kinds in reality because nothing has feature N, if feature N is not really a part of what it is to be a natural kind. And this latter question is at least in part—and perhaps primarily—a philosophical matter. Similarly, there are many other assumptions at play in science that are not empirically grounded, concerning, for instance, the notion of complexity (see Rocca and Andersen 2017). Some accept reductionism, for example. Perhaps one science is reducible to another, such as if we can explain psychology in terms of evolutionary biology or economics in terms of neuroscience (see Chapter 14). Perhaps everything is reducible to physics. Again, though, we need to understand what conditions would qualify as a reduction, which then gives us a grip on what, if anything, would allow us to recognize a case of it in reality.

Our job, however, concerns specifically causation in science. One task is to tease out the assumptions that are being made about the nature of causation: articulate them so that we can scrutinize them and judge whether they are right. But this is not our only task; indeed it is not our main task. This book is primarily about the nature of causal science and its own philosophical presuppositions, as manifested in the norms of causal science. What we mean by causal science is the science of finding causes in the natural world and using that knowledge. We believe that smoking tobacco causes cancer, for example, and that the heating of iron rods causes them to expand. Do we have good grounds for these beliefs? We can see that this comes down to the question of whether the correct norms of causal discovery have been followed and given a positive judgement. But do we know that those correct norms have been yet found? Furthermore, is it possible that some norms of causal discovery are inappropriate? This could again be a partially philosophical matter. Suppose one has a regularity theory of causation: that it consists in an absolute regularity or constant conjunction (Hume 1739: I, iii, 2). One may then adopt as a norm of causal science that one should record all the constant conjunctions and draw causal inferences from them. But this norm is inappropriate, we allege, because it will fail to reliably identify the causal facts. It could pronounce that smoking is not a cause of cancer, because not all smokers get cancer; or, contrariwise, that humanity is the cause of mortality, which we would say it is not, even though all humans are indeed mortal. What these examples illustrate is simply that our philosophical view of the nature of causation shapes the norms that we adopt for causal science; that is, it determines what we think we should look for when seeking causal connections.

1.4 The Critical Friend

The defence of a philosophical contribution to science should in no way be understood as an attack on science. Although science cannot answer every question, it still gives us the best way to answer questions of a certain kind, including the empirical matter of what is causally connected to what. Philosophy should be seen as science's critical friend, pointing out those normative aspects that in part constitute science and examining the credibility of the underlying metaphysics upon which it rests. In return, science can be philosophy's critical friend, too, allowing the abstract theory to be informed by a better understanding of the concrete reality. It might be assumed in the philosophical account, for instance, that there cannot be instantaneous action over a distance. The Einstein–Podolsky–Rosen experiment in physics shows an alleged case where perhaps there could be such action, under one interpretation. The philosopher then has an option of dropping the stricture from her metaphysics of causation. But is the interpretation a sound enough basis for dropping the stricture? There can be, thus, a toing and froing between the concrete and the abstract as our understanding develops. The goal is to find a theory that is satisfactory both scientifically and philosophically. This is what we mean by a reflective equilibrium.

In this book, we offer a sample of philosophically oriented discussions with which scientists should want to engage because they relate to the scientific discovery of causal connections. Should science construct causal theories, or only deal with data? How can our scientific studies have external validity, beyond the data set? And can we assume that the more data, the better is the evidence of causation? What do we mean when we interpret a theory probabilistically? And why do we repeat experiments if they have already been shown to work once?

These questions show how normative philosophical considerations and empirical science can be intertwined and mutually informed. If treated by the philosopher in isolation, we will miss out on important evidence of how the world is observed to work. At the same time, science will be impoverished, lacking critical reflections over its conceptual, ontological, and methodological foundations. Thus, through engaging in metascience, there are good prospects of finding a complete and full understanding.

Uncovering the causal connections within nature is a key role for science. We must therefore consider how well the methods of science are suited for this task. Since all methods seem to reflect a set of deeper conceptual and ontological commitments concerning the nature of causation, we will have to make these commitments explicit and open for scrutiny. This is a philosophical task to which we now turn.

2

Do We Need Causation in Science?

2.1 Causal Scepticism

There is a well-known, often quoted claim that causation has no place in science. We need to address this challenge as a matter of urgency for, if true, there would be no point in us continuing this book. There would be no causation in developed scientific theories for us to investigate. The claim is articulated by Bertrand Russell:

> All philosophers, of every school, imagine that causation is one of the fundamental axioms or postulates of science, yet, oddly enough, in advanced sciences such as gravitational astronomy, the word 'cause' never occurs...The law of causality, I believe, like much that passes muster among philosophers, is a relic of a bygone age, surviving, like the monarchy, only because it is erroneously supposed to do no harm. (Russell 1913: 193)

Russell's view is old but powerful and has retained support. There have been continuing interpretations of science—physics in particular—in which it is free of causation (Heisenberg 1959: 82, Feynman 1967: 147, Healey 1992: 193, Price and Corry 2007b, French 2014: 228).

How can this be so? Russell, and those who followed, argued that when you look at a successful and well-developed science such as physics, you find no claims of causal connection, such as regularities of same cause, same effect. What you find instead are equations that express co-variances of various magnitudes, e.g. $F = ma$ or $F = Gm_1m_2/d^2$. Causation, as philosophers have traditionally conceived of it, has a number of features that we just do not see here. In particular, a causal connection is typically understood as directed and asymmetric. Causes produce their effects, rather than vice versa. Many say that causes must precede their effects in time. But we do not get this feature with equations, which exhibit symmetry instead. The equivalences they state can be read from left to right or right to left. Many sciences use differential equations: not just theoretical physics but also applied sciences such as engineering. Using these, we can calculate backwards as easily as forwards from any configuration. Hence, the future 'determines' the past just as much as the past 'determines' the future (Russell 1913: 201–2). In a science whose central tenets are articulated in the form of equations, therefore, there is no asymmetry represented. Allegedly, then, a physicist has no use of a concept of causation because it would commit us to asymmetric determinations that are just no longer believed by the experts to be features of the world.

What, then, should we say about the status of causation? In particular, how can we account for the common practice of thinking of the world in terms of causal connections? Even in an apparently scientific context we say, for instance, that heating an iron rod made it expand, the Fukushima nuclear power plant disaster was due to a tsunami caused by an earthquake at sea, or that an acceleration of a particular body occurred because of a collision. All of these look, whether explicitly or implicitly, like causal claims.

Those who follow the spirit of Russell's account usually admit that there is still plenty of causal talk, sometimes even by scientists, on informal occasions. But causation nevertheless does not appear in precise statements of scientific laws and should not, therefore, be deemed ultimately real. As causation is not a genuine part of the world, in serious science we should make no causal claims.

Causation, according to this view, could be seen as nothing more than 'folk science', which means 'a crude and poorly grounded imitation of more developed sciences' (Norton 2007: 12). Similarly, Ladyman et al. (2007: 3–4) think that causation is one of the pre-scientific notions that some have attempted to 'domesticate' into modern science, persisting with a common-sense metaphysics that has no basis in contemporary physics. Norton's argument is restricted not just to physics, though. He thinks that no developed science needs the notion of causation:

What I do deny is that the task of science is to find the particular expressions of some fundamental causal principle in the domain of each of the sciences. My argument will be that centuries of failed attempts to formulate a principle of causality, robustly true under the introduction of new scientific theories, have left the notion of causation so plastic that virtually any new science can be made to conform to it. Such a plastic notion fails to restrict possibility and is physically empty. This form of causal skepticism is . . . motivated by taking the content of our mature scientific theories seriously.   (Norton 2007: 12)

Norton contrasts causation with a respectable notion such as energy, which is to be found across a range of sciences. Among the things he says of it is 'there are innumerable processes that convert the energy of one science into the energy of another, affirming that it is all the same stuff. The term is not decorative; it is central to each theory' (Norton 2007: 14). The term causation, we can conclude, is mere decoration.

Can the causal realist respond to this sort of charge and answer the sceptic? We argue that there is a good answer and we should not abandon causation in science. Indeed, our conclusion will be even stronger than that. We argue that causation is vital for science: its bedrock, a pre-condition for its very existence. We proceed as follows. First we point out that the evidence from science far from warrants a firm conclusion that there is no causation. Second, we argue that science makes essential and ineliminable use of the notion of causation, with the centrality of intervention. And, third, empirical science is dependent on a notion of observation such that it would be rendered incoherent unless the reality of causation is accepted. Taken together, these three points add up to a powerful argument in favour of there being causation in science.

2.2 Interpreting Science

First, suppose we restrict ourselves to physics, following the claims of Russell, Price (2007), and others in assuming this to be a science that invokes equations rather than causation. Perhaps this is the science where causation is most at risk; especially so if all other sciences ultimately reduce to physics (see Chapter 14). But here we should note that physics, as stated, is still open to philosophical interpretation, which is a matter of debate. Until such debates are resolved, are we really in a position to jettison so fundamental a concept as causation? Wouldn't we have to be very sure of our understanding of physics before we did so?

Physics is not yet complete and our understanding of it probably even less so. Are there really no asymmetries in reality? Some think the world exhibits a continuous increase in entropy—net energy dispersion—which provides an objective difference between past and future. Whether this is the case is controversial (Kutach 2002, Frisch 2007, Loewer 2007). Similarly, some think that physics allows for the possibility of backwards causation or 'retrocausality' (Price 1996, Price and Weslake 2009), thereby rendering the asymmetry of causation an illusion. Perhaps in some cases, effects can precede their causes, but this is also controversial. Norton (2007: 22) is at pains to point out that, even in Newtonian physics, some events are uncaused. Suppose we accept the claim. It still does not undermine what most causal realists believe. That causation is real does not require that every event has a cause. The realist's view is that causation is a part of reality in the cases where it does occur. All the above issues remain contested.

Another matter is how we are to understand the equations of physics. They look like a mathematical description or modelling of reality. But surely they should not be conflated with that reality. Although contemporary physics makes extensive use of maths, isn't that just a tool for representing certain structural features of the world? The world itself is not an equation, nor a number. Furthermore, an equation may deliberately represent only selected aspects of reality, for purposes of abstraction. A philosophical interpretation—outside of science itself—would then still be needed to supplement that representation in order to give us a complete understanding.

Examples might help. An equation fails to indicate asymmetry but that does not entail that there is none in the relationship between the things represented. Thus, although the formula $F = ma$ is correct, this does not rule out some asymmetric causal relationships. We might causally intervene to increase the acceleration of an object by increasing the force on it, for instance. To take another case, there is an equation relating the length of a man's shadow on flat land, his height, and the position of the sun. From any two of these variables, we can calculate the third. Although this equation is in that respect symmetric, we also understand the direction of the causation involved. The length of the shadow will increase when the sun lowers or if the man stands on a box. But there is no intervention on the length of the shadow that changes the position of the sun. We cannot straightforwardly infer, then, from the failure to represent causality in the equation that there is no causation in the world. It might just be that physics is concerned only with some features of the world—its mathematical structure—but not others.

Finally, on this issue, we can refer back to the naturalistic trap described in Chapter 1. Whether there is causation in science cannot be decided by looking exclusively at science. We also need an understanding of what causation is, and this is at least in part a metaphysical issue. In particular, no one should be allowed to take alleged features from a flawed theory of causation, claim that physics has none of those features, and then conclude that there is no causation. That would count against only 'straw man' causation. Russell (1913), Heisenberg (1959: 169), Kutach (2007), and Norton (2007) are all sceptical of causation because they think science has no use for a notion of determination or necessity between cause and effect, for instance. But not all theories of causation are committed to this (Mumford and Anjum 2011, Andersen et al. 2018). It depends on the question of whether causes produce their effects by necessitating them. To settle this, we need an abstract, metaphysical consideration of the nature of causation itself. Someone with a firm commitment to naturalism will decline to engage in this kind of metaphysical study. It is clear that they are working with some conception of causation, however. And if it's an imperfect one, then that is a bad reason to conclude there is no causation in science.

The argument of this section is inconclusive. That is exactly our point. Merely looking at physics, without its interpretation, is inconclusive on the matter of whether there is causation in science. However, we will now move on to arguments that seem more decisive.

2.3 How Could We Experiment without Causation?

Our first argument commences with the observation that an important part of science concerns experimental intervention: changing one thing in order to change another, or to see what changes. This looks like causation. How can one thing change another unless there is a causal action?

Now it may be possible to claim that this appearance is superficial and can be cashed out in other terms. Similarly, Mach (1960: 325) invoked a notion of functional dependence, which he thought could replace our primitive notion of causation. But can this view withstand scrutiny? In many plausible cases, how one magnitude functionally depends on another looks to be a matter of causation (not in all cases: the volume of a cube is non-causally a function of the length of one of its edges). Consider the man and his shadow, again. Is it just pure chance—a brute fact—that the length of the shadow is a function of the height of the man and position of the sun? That would seem remarkable—indeed miraculous—given how that exact function holds for such a range of variables. Fortunately, there is a convincing explanation of why the length of the shadow is a function of the other two phenomena: namely that the man's body is causing the shadow by blocking the light of the sun. Causation explains the equation and thus the functional dependence.

It looks, therefore, as though science has a causal commitment in the idea that changing one thing changes another. The point here is that the very idea of an experimental intervention depends on this. Consider the case of CERN in Geneva. This research project aims to smash atoms apart using a process of particle acceleration and collision, from which we can find the inner structure, if any, of some of the smallest constituents of matter. The project is believed to have verified the existence of the Higgs boson. The Large Hadron Collider is the main facility or tool of CERN. It is a tunnelled ring encircling Geneva, 100 metres underground. Setting aside the start-up costs, CERN's budget was 1118.3 million CHF (approximately £767 million) in 2015 alone. This raises a simple question. If we do not believe in causation—and if physicists in particular did not believe in causation—why are we giving them all this money to do their experiments? If the Large Hadron Collider is not a way of intervening, in this case by smashing together accelerated proton particle beams, why is it worth the money? Unless the experiments cause changes, they are surely not worth performing or funding.[1]

Now there are a number of deflationary views of causation, some of which have been attractive to philosophers of physics. A deflationary account allows us still to talk of causation, perhaps usefully and truthfully, but tells us that causation is much less than commonly believed. Price and Corry thus support a position that they call causal republicanism, which is that 'although the notion of causation is useful, perhaps indispensable, in our dealings with the world, it is a category provided neither by God nor by physics, but rather constructed by us' (Price and Corry 2007a: 2). Similarly, Norton thinks our causal explanations are merely pragmatic. Reiss (2012) sees causation as fundamentally subjective, playing a role in human reasoning practices; yet it is partly objective, since predictions and interventions based on those inferential practices can be objectively evaluated. Beebee (2007) advocates a projectivist account, where we project causation on to the world, when really it comes from our own human nature. Menzies' (2007) perspectival realism and Price's (2007) causal perspectivalism have similar features, in which causation resides only in a particular point of view on the world. Could it be that the causation in physics, and elsewhere, admits such a deflationary interpretation?

But then wouldn't we be disappointed in the project of science if this really were the case? Wouldn't we still be right to question the level of investment in experimentation? Suppose we believe that CERN doesn't really intervene: it doesn't really collide particles, it doesn't even accelerate them. Rather, such scientists are engaged in a social

---

[1] Thanks to Fredrik Andersen for making us aware of this point, and some of those immediately following.

activity, created by us, in which, from their point of view, it looks like they are doing something that in reality we all concede they are not. If causation is really not part of physics, then that CERN budget looks like a very expensive job-creation scheme to give the scientists something to do with their time: an extravagant social welfare project.

We do not imagine for one second that this is how it is. Science and experimentation are worth it precisely because they involve causal interventions in the world. Every scientist must believe this.

2.4 How Would There Be Any Data without Causation?

In his defence of energy in science, Norton accepts that energies are measurable. This is a good reason to accept something as real: that it is measurable. But how can something be measurable unless there is causation? Suppose you want to test the energy level in a car battery. You attach a voltmeter to it and see if the hand moves across the dial and, if so, you take a reading. The question is how this would count as a measurement of the energy in the battery unless the battery's stored energy was causing the movement of the hand across the dial. If the hand was moving randomly, all the time, whether the meter was connected to a battery or not, we would not think it was measuring the energy of the battery at all. To count as a measurement is to be causally affected by the thing measured, and proportionally so. Not every measurement requires such causal affectation—holding a ruler against something doesn't— but many cases do. An artefact counts as a meter, such as a voltage meter, only because it is appropriately affected. Similarly, devices such as Geiger counters and oscilloscopes are telling us information about the world—reading something off from it—only if they are being causally affected by it in some way. Their clicks or wavy lines would otherwise be meaningless.

We can see that what we say of measuring devices can be said of observation generally, not just when it is observation performed via some apparatus. Even to make an observation with the naked eye entails being causally affected by the thing observed, otherwise it would not count as an observation of that thing. It would otherwise simply be imagining or misperceiving that thing (Anjum and Mumford 2018a: ch. 6). For anything to count as data, therefore, there has to be some causation. The intervention side of science is meaningless without causation; and we see now that the same applies to the observation side. Hence, the scientist needs causation to be able to perform an experiment but also to be able to find a recordable result.

Russell himself came to see this. Most who cite his 1913 scepticism about causation (e.g. all those in Price and Corry 2007a), neglect to mention that Russell changed his mind about the role of causation in science. Instead of causation being dispensable or even harmful to science, by 1948 Russell thought of causation as one of the fundamental postulates of science: part of its foundations. Its indispensability for the possibility of observation was just one case where it had to be assumed. Writing about the empiricist philosopher David Hume, whose scepticism about causation as anything more than mere regularity has proven enormously influential, Russell noted:

Hume, in spite of his desire to be sceptical, allows himself, from the start, to use the word 'impression'. An 'impression' should be something that presses in on one, which is a purely causal conception. The difference between an 'impression' and an 'idea' should be that the former, but not the latter, has a proximate cause which is external. (Russell 1948: 473)

Hume's use of impression corresponds to our notion of perception or observation. Russell's argument is clear. Without causation, the chief commitment of empiricist philosophy—that all knowledge comes from experience—doesn't even get off the ground: for how would we gain knowledge of the world unless the world was the cause of our experience?

2.5 Limits of Empiricism

Hume's views have shaped the subsequent philosophy of causation in profound ways and we think this has not always been helpful. Hume (1739, 1748) would not allow as legitimate any idea that could not be traced back to some original impression or observation. Many believed causation to be some real connection, over and above the regular and predictable sequence of events that we see. This was a mistake, he insisted. No such connection was perceivable. We could witness one event followed by another but we only believe that the first causes the second because we have seen many cases of the first type of event and they have always been followed by the second type. To go beyond the observable is to enter into metaphysics, which Hume thought ultimately meaningless. Such empiricism is a major motivation behind philosophical naturalism, of the type discussed in Chapter 1. Indeed, Humean ideas were deeply engrained in logical positivism (see Hanfling 1981) and its attack on metaphysics.

One can see that if such a position is adopted, causation becomes an immediate object of suspicion. It can be no more than what is evident from the data, and then the business of science should be restricted to the data: for instance, recording correlations. So the empiricist philosophical position has implications for how we discover causes in science. But if the later Russell is right, this approach collapses. There are thus limits to empiricism (Russell 1948: VI, x). It requires some non-empirical postulates to get started, most of which involve appeal to the reality of a causation that cannot be known through experience. Nothing counts as experience of the world, after all, unless it has been caused by it. There would be no data without causation.

2.6 The Tasks of Science

Causation is indispensable in science. The avoidance of causal talk seems insincere and, we will argue, unnecessary. Its role in science is more than merely incidental. We defend the centrality of causation in science (see also Illari et al. 2011). Scientific methodology explains how causal theories are generated, evaluated, and put to use. We argue that the main purpose of science in general is the discovery of causes and use of them in explanation, prediction, and technical application. These are all inherently causal matters, and without them, science seems to have lost much of its point.

Once we have discovered a case of causation, it gives us a power to get what we want. We spoke of experimentation as involving an intervention in the world but the idea of an intervention is even more important once our experiments are complete. In medical science, for instance, if we have found that some drug causes alleviation from suffering, then it allows us to intervene in a particular case in the expectation of achieving that effect. So the finding of causes is one very important task.

We do not have to say that the whole of science is directly about discovering causal connections, however. Science has other tasks, such as identifying what exists: what there is. Science tells us that there are electrons, leptons, what is the number of married men, or the average age of a population. Further, science aims also to find the essence or structure of something, such as an atom or a gene. It may have other jobs besides that are not explicitly part of causal science. But even these, we claim, are likely to depend on causation in the end. What exists, for instance, might come down to whether something makes a difference to the world. Electrons exist if they can cause something else to happen or enter into interaction with other things. And the structure of something could be understood as a causal structure involving components suitably arranged.

We have not yet even discussed the idea Russell developed in his later thinking that causation could provide some grounds for inductive inference; from the observed to the unobserved. We think we have already said enough, however, to justify a stance that we will take forward to the rest of the book. This is that causation is fundamental to science and cannot be eliminated from it. Some interpreters may have been particularly bad at spotting causation in science, invoking notions such as intervention and measurement without understanding them to be dependent on causation. We have seen some grounds to argue that the whole practice of science would be worthless unless there was causation. This conclusion established, we can, after all, proceed to the main purpose of the book, which concerns how science goes about discovering causes.

3

Evidence of Causation Is Not Causation

3.1 A Crucial Distinction

We have argued that if we want to discover causes, we first need to know what it is we are looking for. A distinction needs to be drawn, therefore, between two related issues: what causation is, and how we come to know of it. Although we urge that these are two separate matters, they are clearly connected: what we say about one will affect what we ought to say about the other. The first question—what is causation?—is an ontological matter concerning the nature of causation itself. To understand the nature of causation, it is useful to know, for instance, whether causes necessitate their effects. Are causes prior to effects or can they exist simultaneously? Is causation a process or a relation? Do effects counterfactually depend upon their causes? To understand what causation *is*, we need to settle these and other ontological issues. But the question of what is the best way to *discover* causation is a matter of epistemology or theory of knowledge. Scientists engage in an epistemological debate when they discuss which research methods are best suited for discovering causes, but their answers will be informed by what they think causation is, ontologically.

Why is it so important that ontology and epistemology be kept apart when clearly they are closely related? We can, however, see that there are good reasons to keep knowledge of a thing separate from the thing itself. The distinction between ontology and epistemology is essential in a court of law, for instance. In principle, there could be such a thing as a perfect crime: where the crime is committed but there is no evidence of it. A murderer could get rid of the body, remove all traces of DNA from the crime scene and make sure that there were no witnesses. But that there is no evidence of a crime does not mean that no crime was committed. Legally, one cannot judge anyone guilty of a crime unless there is sufficient evidence that the accused actually committed it. So even if the defendant committed the crime, one needs evidence to gain knowledge of it.

Science seems no different from law in this respect. A scientist must be careful not to make causal claims in their research without evidence to support them. At least in this sense, ontological claims depend on some epistemology. What we think there is depends on what evidence we have. The reverse is less acknowledged but still a case can be made for it. Epistemology depends on ontology. When we search for causation, we do this with a preconception of what we are looking for. This preconception is rarely explicit, or even reflected upon, but it will unavoidably influence what we can find.

Consider the legal case again. When searching for a murderer, our focus will be on finding and interpreting the evidence. But what counts as relevant evidence is specific: people seen in the area at the time of the murder, witness reports, traces of DNA, footprints, and so on. In order to keep a focus in the search, it is essential to filter out irrelevant information. Only information that relates to the murder will count as evidence of it, and to know what counts as evidence of murder, one must know what a murder is. In other words, one can take something as evidence of murder incorrectly.

Similarly, if the scientist is wrong about the real nature of causation, she might be looking in the wrong places or even for the wrong type of thing. Now it might be easier for everyone if some suitable expert could simply let the scientists know the ontological truth about causation so that they could always be sure they were looking for the right sort of thing. A problem is that philosophers disagree on the nature of causation and thus also disagree about what the scientist should be looking for. Should it be perfect regularities (Psillos 2002), probability raisers (Suppes 1970), counterfactual dependences (Lewis 1973b), manipulability (Woodward 2003), INUS conditions (Mackie 1980), processes (Salmon 1984, 1998, Dowe 2000), mechanisms (Glennan 1996), causal powers (Cartwright 1999, Mumford and Anjum 2011), or what? Also within science there are diverging views about causation, and what is considered to be the best evidence of causation varies from one discipline to another. What we see, then, is that there is no general agreement over what causation is within philosophy or within science. Nevertheless, a number of ontological assumptions about the nature of causation are made in science, we contend, simply through the choice of research method.

3.2 Evidence Is Restricted by Methods

How we go about searching for causation reveals something about what we expect causation to be. One might say that epistemology, in this sense at least, carries with it ontological commitments. What counts as evidence of causation depends on what types of results can be produced by our methods, which again depends on which methods are available, accepted, and promoted by the scientific community.

Science relies heavily on methods of investigation. With the scientific revolution, science became increasingly concerned with methods and methodology as a way of ensuring suitable rigour in its practices. In his work, Galileo (1632) introduced the use of experiments, idealizations, models, hypotheses, and mathematical tools. This methodological framework is now seen as an integrated part of science, and to such a degree that it has become almost definitive of it. To separate science strictly from its methods might not be possible, and perhaps not even desirable, if its methods are in part constitutive of science. Gower suggests that science and methods must be understood as mutually dependent: '[Scientific beliefs] have something in common which explains why we count them as scientific, namely the method by which they are established' (Gower 1997: 6–7). Unless the ways in which we go about collecting data, performing experiments, and generating results are controlled, it would be difficult to separate science from any other form of activity, or what Popper (1962) called pseudo-science. Without a clear methodology, therefore, scientific evidence seems to be up for grabs.

We are still far from reaching that stage. Science is trusted primarily because of the way in which it arrives at explanations and predictions, and pseudo-science is distrusted for the same reasons. Suppose a psychic predicts a warm and dry summer and a meteorologist does the same. Unless we accept the psychic's methods as scientific, we will not call his prediction scientific either, even if it turns out to be correct. Likewise, if we accept the meteorologist's methods as scientific, we will not say that her prediction was non-scientific even if it turns out to be wrong. It seems, then, that being deemed scientific is to a large extent a status bestowed when the right methods are used.

Note that there is a difference between methods and methodologies. Methods are the actual procedures we undergo in performing experiments. They are what one has to learn already as an undergraduate student. In statistics, for instance, there are a number of different methods (e.g. data collection, data summarization, statistical analysis), and not many scientists will master all methods equally well. A methodology, on the other hand, is an overarching view of how science should proceed, such as whether the experiments performed are based on inductivist or falsificationist principles, or others. Few researchers will ever find themselves concerned with these principles, although they will undoubtedly be part of a scientific practice that favours one methodology over another. One could be a pluralist about methods, arguing that there are many methods that are suitable for establishing causation, while advocating a single methodology. One could be a pluralist about both methods and methodology, or a monist about both (see Chapter 27 for this discussion).

Scientific methods are not always treated as equal, however. Some methods could provide better causal evidence than others, and there is a more or less explicit expectation that scientists should choose the methods that rank highest on the evidential hierarchy. In economics, statistical tools and quantitative methods are typically preferred to experimental methods, while in medicine, randomized controlled trials and meta-analyses of these are thought to outrank both correlation data and any type of qualitative methods, at least when searching for causation. In chemistry and physics, however, priority is usually given to experimental methods and lab settings over purely quantitative studies. Still, statistical methods can be used for certain purposes also in physics, for instance to study measurement errors of random phenomena and for approximate results.

What are the philosophical implications of the methodological divergences in science? First of all, it seems clear that scientific evidence of causation does not mean the same across disciplines. Different methods provide different types of evidence. So even if we agree that policies, practice, or decisions should be based on evidence, there is the further question of exactly what type of evidence we should consider. This is important because evidence is shaped and restricted by our choice of methods.

Suppose we want to find out whether there is a causal link between gender and teaching scores for academics. The type of knowledge we can get through a questionnaire is not the same as we can get through an experiment. A survey or interview should typically only record information that is already there, without influencing it, while in an experiment one can manipulate a situation to provoke an effect that normally would not happen without such intervention. A survey might then show that female teachers get lower scores than male teachers, but it would leave open whether this is due to gender bias or to the quality of teaching. In an experiment performed by MacNell et al. (2014), in contrast, this was tested by using online classes to hide the teachers' gender from the students, which allowed each teacher to operate under one male and one female identity. When the students then gave significantly better scores to the male version of the teacher than to the female one, it suggested that the difference in teaching scores was caused by gender bias and not by the teaching quality.

This leads us to a second implication of the methodological divergence in science: that we cannot treat evidence as an ontologically neutral matter. We cannot say that statistical evidence is the same as evidence from single case studies, but only on a larger scale, for instance. The two types of evidence are of distinct kinds. In quantitative studies, we gather data from a large number of cases, while in qualitative studies there is typically more data from fewer cases. What do these different methods tell us about the ontological commitments of the scientist as regards the nature of causation? That it shows itself in repeated patterns? That it can be found in the details of a few cases, or even in only one instance?

What we accept as evidence of causation thus relies heavily on the choice of methods. In fact, each scientific method is developed to latch on to a certain purported feature of causation. Which method we give highest epistemological status will therefore depend not only on the discipline, but also on the preferred theory of causation. Only when our choice of method matches the correct understanding of the nature of causation, can we claim that the evidence generated by the method is evidence of causation. In other words, our epistemological approach must fit our ontological commitments.

3.3 Causation Is Not Restricted by Evidence

Even if evidence and scientific knowledge is constrained by our methods, the same does not hold for causation. Evidence of causation is not constitutive of causation.

We must therefore be careful not to mistake the tool of investigation, the method, with the subject of investigation, causation. That would be to collapse ontology into epistemology, which we argue must be kept apart. But not everyone agrees with this sharp division.

Philosophical idealists such as Berkeley (1710) say that the only evidence we can have for the existence of material objects is our perceptions of them. From this, Berkeley concluded that material objects are therefore nothing but perceptions. We think that the conclusion does not follow, but instead is an example of what Dyke (2008) calls a representational fallacy, or what Whitehead (1929) refers to as the fallacy of misplaced concreteness. Similarly, it is tempting to argue that what we know of causation is only what Hume thought—a constant conjunction between two types of event—and therefore causation is nothing but such constant conjunction. This we also take to be a representational fallacy, namely, mistaking the evidence we have for a thing with the thing itself.

To identify causation with how we come to find out about it is a form of operationalism, which is to identify a phenomenon with the operation we would have to perform to verify its presence. One might identify time with the ticking of a clock or temperature with the reading on a thermometer, for instance. We reject these identifications. Nevertheless, there are some parallels between the operationalization of causation and other such identity claims in science, and looking closer at one of these might make it easier to understand why operationalization is problematic in general.

Physicist and Nobel Prize winner Percy Williams Bridgman noted that scientists often use very different types of measurement tools for what was assumed to be the same type of phenomenon (Chang 2009). For instance, when measuring distances to other planets, one does so in light years, but when measuring distances between two houses, one could use a rod. These two types of measurement seem, he argued, to rely on very different concepts of distance. 'To say that a certain star is 10^5 light years distant is actually and conceptually an entire different *kind* of thing from saying that a certain goal post is 100 meters distant' (Bridgman 1927: 17, 18). So instead of saying that we in both cases measure length, he said that we should specify exactly what it is that we are doing, so as to avoid confusing different concepts: 'In *principle* the operations by which length is measured should be *uniquely* specified. If we have more than one set of operations, we have more than one concept, and strictly there should be a separate name to correspond to each different set of operations' (Bridgman 1927: 10). What seems to be at stake here, is the nature of the phenomenon that we are studying through different methods. If we are not specific about our choice of operational tools, Bridgman said, 'we may get into the sloppy habit of using one word for all sorts of different situations (without even checking for the required convergence in the overlapping domains)' (Bridgman 1959: 75).

The same caution could apply to causation. Different types of evidence of x might actually involve completely different interpretations of what x is. Such a conclusion may tempt us to causal pluralism: the view that causation is one word but many different things in reality. But the attraction of operationalism effectively depends on ignoring the distinction between what we can establish evidentially and what it is in the world ontologically. This is a critical mistake especially when dealing with causation, as we will try to explain in following chapters. This book is a defence of the ontology/epistemology distinction in the causal sciences.

3.4 Is Causal Evidence Based on Causation?

We are now ready to look closer at the relationship between the methods we use for establishing causation and what we think causation is. We can start with Hume's empiricism, which affected his view on how causation is to be understood. Empiricism is the view that all we can legitimately know must ultimately have its origin in what is experienced through our senses. An empiricist philosophy of science will then urge us to adopt scientific methods that deal only with empirically observable data and their analysis. Since Hume thought that causation cannot be known beyond the observation of constant conjunction, temporal priority and contiguity, he concluded that we have no reason to believe that there is anything to causation over and above this (Hume 1739). Direct observation of causation then becomes a problem for those who believe that causation involves something more, such as a necessary connection (e.g. Aristotle *Metaphysics* Θ 5: 264, Spinoza 1677: I, axiom III: 46, Kant 1781: II.ii.3 second analogy, Mill 1843: III.v.6, Mackie 1980: 62) or causal powers (Harré and Madden 1975, Mumford 1998, Molnar 2003, Cartwright 1999, Martin 2008, Mumford and Anjum 2011). Necessary connections and causal powers are allegedly not observable, whereas constant conjunction, temporal priority, and contiguity supposedly are.

If causation is something that cannot be immediately observed, it seems that we can only know it by inference from something else that is observable, but therefore also not the same as causation. Scientific methods might help us find certain features from which causation could then be inferred, under specific conditions. Observation studies require us to look for robust correlations, while comparative studies such as randomized controlled trials are more suitable for finding difference-makers, for instance. Experimental methods seem to rely upon the manipulability of causes and statistical methods work well to find raised probabilities. By searching for causation via any of these features, we are essentially accepting them as reliable indications of causation.

Causation should not be identified with the successful outcomes of such studies, however, since that would be to mistake causation with evidence of causation (Anjum et al. 2015). Instead, if we want to find the best method for establishing causation, we should first ask what causation is. Only then will we have a reason to say that one method is more suitable for finding causes than another. If causation actually is a difference-maker, for instance, then comparative and contrasting methods may be the best way to detect it. But if causation is instead a constant conjunction of two types of events, we seem better justified in using large-scale correlation data. Should causation turn out to be something else entirely, then what would be the rationale behind searching for statistical correlations to find causes?

We will argue that none of the proposed characteristics of causation are immune to counterexamples, so they cannot be taken as identical to causation, nor as entirely reliable markers. In the course of this book, we will see that not all causes make a difference (Chapter 15) and that not all causes can be manipulated (Chapter 24), which is why we must sometimes use thought experiments rather than physical experiments. We will also see that some necessary conditions cannot plausibly be taken as causes, at least not without also accepting the Big Bang as the cause of everything that has happened since (Chapter 17). And even some of the most robust correlations turn out to be non-causal (Chapter 4).

As noted by Bridgman with his example of measuring length with different tools, a consequence of having plural methods is that they might not even pick out the same thing. The same can be said about causation (Kerry et al. 2012). If we have evidence from more than one type of method, we might end up with conflicting results: some suggesting the presence of causation and some suggesting no causation. So which method should we trust more? We will argue (Chapter 10) that many of the observable features that scientists can find through their methods are merely symptomatic of causation, rather than constitutive of it. None of these symptoms are definitive of causation, meaning that there can be causation without them. What we are left with, then, is evidence of exactly what we have been looking for: correlations, difference-making, probability raisers, and so on. But would we even be interested in these types of evidence if we didn't think that they were somehow indicative of causation?

3.5 Ontology First

We started this chapter by proposing that a distinction be drawn between epistemology and ontology and it should now be clearer why. Our methods pick out certain features or symptoms of causation, and different methods look for different features. If we then mistake the symptoms of causation (epistemology) for causation (ontology), we will get different notions of causation from using different methods. And while some might be happy with a pluralistic notion of causation (e.g. Hall 1994), we will also see that it has some problems and that there are other available interpretations of the same facts. Hence, one possibility, which we will endorse, is that there are plural evidential approaches to the causal facts but that causation is one single thing; that is, an evidential pluralism but ontological monism concerning the nature of causation (see Chapter 27).

It has also been seen that once a distinction is established, between ontology and epistemology, there can also be a question of priority. With which should we begin? Does one dictate the other? Here we have offered a conceptual and methodological argument: that without a prior conception of causation, one cannot know what to look for as evidence of it. The upshot of this argument is that ontology has priority. Empiricism offered a different view: that we must always construct our concepts from experience, which is thus to take epistemology first. In the case of causation, we have already presented (in Chapter 2) a major difficulty for this view. The significance of experience is itself underwritten by the reality of causation, in which case a project of 'constructing' causation out of experienced events seems misplaced. This becomes another reason, therefore, to put ontology before epistemology.

In this first part of the book, we have offered justification for the most basic commitments that shape the rest of what follows. We have argued that philosophy has a valid contribution to make toward the empirical sciences (Chapter 1). We then argued that causation is essential to science and cannot be eliminated or reduced away on the supposed grounds of naturalism (Chapter 2). Then, here, we have drawn a distinction that, when applied to our topic, means that we should keep causation as it is in the world separated from the evidence we gather in our attempt to confirm its presence. With this groundwork complete, we can now proceed to examine the issues and problems raised by various approaches and methods for finding causation in science. We will see how those methods sometimes disappoint us when they fail to get the nature of causation right.

PART II

Perfect Correlation

4

What's in a Correlation?

4.1 Regularity, Alive and Well

It might be thought that there is no point attacking the regularity view of causation. It is a theory that virtually no one supports openly anymore, in either philosophy or science. Why waste time with a straw man, then? Hume (1739) produced this view, arguing that there was nothing more to causation than a regularity or constant conjunction. We think that A causes B just because A is regularly followed by B, he said. Hume thought there were just two other conditions that had to be satisfied: causes and effects were next to each other spatially and the cause preceded the effect in time. Surely we have made considerable progress since then.

It is premature to announce the death of the regularity theory, however. In this part, we argue that it is alive and well in many of the ways we think about the world and conduct science. If the theory is wrong, which we believe it is, then we still have to rethink our scientific approach. There is a danger in assuming the regularity theory to be banished, when really, we will say, it has only been transformed, developed, and exists now embedded within our empirical practices. This is reflected in the importance we still attach to correlation data when searching for causes. In this second part of the book, we will consider how philosophical thinking about causation as regularity is reflected in widely accepted methods for discovery of causes.

4.2 Regularity as the Starting Point of Science

Scientific theories involve a complex combination of reasoning and experience. There are rational constraints on what makes a good theory. It should not involve a contradiction, for example. But logic alone cannot give us a full account of the natural world, unaided by experience of what is known to happen. We need empirical evidence for our theories, delivered by direct or indirect observation of the world.

This norm of science has a defensible basis, which can be illustrated particularly well in the case of causation. Using logic alone, you cannot simply deduce what is causally connected to what, prior to experience (a priori). Hume made this point well (1748: I, iv, 1, 27). The apocryphal first man, Adam, at the initial moment he encountered water, could not deduce that it would quench his thirst or that it could potentially suffocate him. He needed experience to show him that.

Once such an anti-rationalist argument is accepted—rationalism being used here to mean that view that reason alone could deliver all knowledge—we see that much of science will come down to the use of empirical evidence. It then becomes vital that we understand what counts as good empirical evidence for the existence of a causal connection.

We should for a moment consider the size of the task of finding causal connections, however, so that we appreciate how important the role of evidence is. We can consider the relative rarity of the causal connection, for instance. This rarity means that if we take any two variables at random then they are far more likely to be causally unconnected than connected. One could choose, for instance, the average length of wild salmon in Norway and the extent of comic book sales in Belgium. We take it that there is no direct causal connection between these two factors. The same will probably follow for any two other random factors. It is quite rare that we would chance upon two variables between which there is a causal connection.

Another way of understanding the same point is to take some previously unknown substance and think of all the possible things that it could do. It might shatter, corrode a class of other substances, nourish animals upon consumption, bond with sulphur, cause or cure cancer. There are infinite possibilities. Only a small sub-set of these possibilities will be the ones that the mystery substance actually can bring about. We therefore have to investigate the world carefully to find out which. Hume put his Adam in exactly this situation in relation to knowing the causal properties of water.

Such considerations provide the context against which we need scientific methods of empirical investigation. A very natural starting point is the idea that we should look first for correlations as these provide good *prima facie* evidence of how the world works; specifically of what is causally connected with what. There is a correlation between the burning of coal and heat, for instance, which gives us a very good reason to think that burning coal causes it to give off heat. In contrast, the burning of coal does not correlate with it dissolving, doubling in size, the user catching a cold, or the Republican Party winning the presidency. There is thus no evidence for any of those latter causal connections.

4.3 What Correlation Might Mean

Both correlation and dependence are precisely defined in statistics; but it is not this technical sense that we will usually mean here when we invoke the idea of correlation as evidence of causation. In more common usage, correlation just means a correspondence, a matching, or mutuality. There are, then, at least five things that we might have in mind when we say we have found a correlation in science. All these are admissible, we argue, because they involve some kind of empirical regularity.

First there is *co-variance*. This is where two variables increase and decrease together, so that when the rate or magnitude of one is high, so is the other, and the same for low values. We might find a large data set, for example, that reveals that the rate at which a university's academic papers are cited co-varies with that university's extent of international co-authorship. One can summarize this co-variance in the claim that internationally co-authored papers attract more citations.

Such co-variance might be thought of as the core case of correlation as it is often what is meant by it in scientific studies. One can say that air pollution correlates with cancer, for example, if there is good evidence that areas with high levels of air pollution occur with a higher incidence of cancers than do areas with low levels of air pollution. Statisticians can measure the degree of correlation and give it a numerical value. This basically specifies how close the co-variance is. One could think of it in terms of plotting the changing values of both variables on a chart and then seeing how close the two curves correspond.

Second, there is plain *regularity*. In its simple form, this means that one thing always follows another. Every human being who has ever been within 50 metres of the 'Elephant's Foot', the solidified corium remains of the Chernobyl plant's melted reactor core, has died within five years, for example. It can then be said that visiting the Elephant's Foot correlates with death. This notion of correlation corresponds very closely with Hume's view of regularity.

Third, we can think of *stable proportion* as a correlation, and see that this is another form of regularity. This is where one factor, A, is not always followed by another B, as above, but where a stable proportion of As are followed by Bs.

Suppose that we find in a data set that one in 1000 contraceptive pill users have developed thrombosis. Let us assume that this is a higher rate than in the case of non-pill users. Now if we examined other studies and trials and found that this raised incidence of thrombosis remained about the same, roughly one in 1000 users, then we can claim to have found some correlation. The stability of this proportion over a number of studies tells us that this does not look to be a random or one-off finding, and thus that there could be something that needs an explanation. It might be classed as a weak correlation—nevertheless a serious one, as it concerns health—because the proportion is low. However, if that proportion remains stable over many instances then it could be considered a strong regularity. Hume thought of these cases as imperfect regularities or, as he puts it in one place, 'inferior degrees of steadiness' (1739, I, iii, 12, 133).

Fourth, we have *invariance*. An invariance is a stable relationship that can be seen to endure through changes in the values of the variables. Invariance can be expressed mathematically in a formula, such as in Coulomb's law $F = k_e q_1 q_2 / r^2$. Here, the magnitudes of electrostatic forces q_1 and q_2 and distance, r, stand always in the invariant relationship expressed in the formula, no matter how those values alter. The invariance can be understood as exhibiting a kind of stable regularity through change.

Fifth, there are *constants*. Technically, a constant is not a correlation since correlations require at least two variables, whereas constants are singular. However, we include constants here because they are an important form of stable regularity. From our observation of what happens in the world, no doubt assisted by some theory, we could find that in a vacuum the speed of light is nearly 300,000 km per second. This is constant, though light moves slower through different mediums, which we can specify. There are other constants in science, such as the gravitational constant, G, and Coulomb's constant, k_e, in the above law, which are again understood to remain always stable. Although there can be good empirical evidence for the holding of such constants, there are clearly questions to answer regarding their evidential basis. They apply in every single case; but how, it might be wondered, can we know that when we haven't and couldn't observe every case? The reason constants appear here, however, is simply that they show some aspect of a relationship that is absolutely regular.

4.4 What Should We Conclude from Correlation?

Assuming that we have evidence of correlation, there is a further question of what we should conclude from it, if anything. Some say that correlation does not equal causation; but there is another view that if you follow the correct procedures and find exactly the right kind of correlation, then you are in a position to draw a causal conclusion (see Aldrich 1995 for a historical overview of correlation and causation).

There are the following possibilities to consider. Suppose A correlates with B, then it could be that:

1. A is a cause of B.
2. B is a cause of A.
3. A and B have no direct causal connection but may have a common cause.
4. A causes B and B causes A.
5. There is no causal connection between A and B.

These need a little more explanation.

Possibilities 1 and 2 are crucial to distinguish. The issue is that some senses of correlation, such as co-variance, have no direction. We may say simply that A and B correlate and it doesn't matter which of A and B comes first in the statement. Causation does have a direction, however, which is why these can be distinguished as two distinct causal possibilities. Umbrella usage correlates with incidence of rain, for example, but there is a clear direction. Hence, we cannot make it rain simply by all putting up our umbrellas. It is the rain that makes us put up umbrellas, rather than vice versa.

The following example illustrates the third case, of common cause. Suppose a correlation is noted between citation rate and international co-authorship. One might jump to the conclusion that adding a co-author from another country to a paper will get it more citations than otherwise. This could be a mistake, however. A plausible explanation of the correlation is that better researchers get more citations and get more people wanting to co-author with them. In such a case, the high quality of the work produced explains both the citation and co-authorship rates and why they will tend to rise and fall together.

The fourth case is one that is seldom acknowledged. The reason for this is that many see causation as essentially involving an asymmetry, which means that if A is a cause of B, then B cannot be a cause of A. But consider the correlation between insomnia and anxiety. These could form a self-reinforcing vicious circle. Someone gets anxious, which makes them sleepless, which makes them more anxious, which makes them more sleepless, and so on. It is not clear in such a case that one factor in the correlation explains the other exclusively.

Now some might say that the case doesn't really show symmetry in causation because it is about distinct episodes of the conditions. Perhaps anxiety at time t_1 causes sleeplessness at time t_2, which causes another instance of anxiety at t_3, which causes more insomnia at t_4. But there are other cases of symmetry that cannot plausibly be explained away like that. Consider two playing cards propping each other up, standing together in an inverted V shape. Card A is causing card B to stand at an angle, and card B is causing card A to stand at an angle, so this seems a case of symmetry. There is a direction to causation but it is mutual: from A to B and from B to A. The case is controversial because it rests also on a thesis that causes can be simultaneous with their effects, though we think such a view can be defended (Mumford and Anjum 2011: ch. 5).

The fifth case is also a serious one to consider: where the correlation is accidental, coincidental, or spurious, in that there is no causal relationship involved whatsoever between A and B. This might be thought to raise the question of whether there is enough evidence available to show the true picture. Perhaps accidents only look like causes due to insufficient data. A coincidence seems more likely with a smaller sample size, one might think. If only three things are A, and they are all B, then this might be mere coincidence. But suppose a thousand things that are A are also all B? Could that really be a mere coincidence? A challenge to such a view is that there have been found to be some very strong correlations based on large data sets but which we nevertheless believe to be entirely coincidental, for example between the per capita margarine consumption in the US and the divorce rate in Maine, which is 98.9 per cent correlated (see Vigen 2015: 7).

If A and B correlate, therefore, there is still work to be done to discover the causal connection that is involved, if any. How can we do this in a way that uncovers the true causal structure and rules out the accidents? We basically have three options. The first is to get more data, and perhaps a specific type of data that we think is of a high enough standard to draw the correct causal conclusions. The second option is to perform an intervention on one of the variables to see whether the other changes with it. This is not always practical or permissible, however. The third option is to seek a mechanistic understanding of the case so we can understand a credible route by which one thing causes the other. This might not be available, as detailed empirical knowledge and theoretical insight could be lacking.

Here, we will just consider the first option, of acquiring more data, and come back to the other options later in the book.

We might first look at an informal approach, such as that of Sir Austin Bradford Hill. We say this is informal because Bradford Hill offered nine criteria for the identification of a cause but he was clear that none of these should be applied 'hard-and-fast' (Bradford Hill 1965). They were more heuristics, or rules of thumb, to be taken into consideration when forming a judgement. We will argue later that this sort of informal approach is the right way in which to appraise evidence, treating it as symptomatic of causation rather than constitutive of it (see Chapters 10 and 27).

Of Bradford Hill's nine criteria, four of them concern the gathering of data that is amenable to statistical analysis. We will consider the others in due course. But we should note here that he urges us to look at the strength of the association between A and B, consistency of the association over different contexts, whether the effect is specific to a particular group, and what he calls biological gradient, which is what we called co-variance.

This gives us the idea that we should not merely passively record any old data that we receive. We should, instead, go out and seek data of certain specific kinds. It seems possible also that we could take a phased approach to data gathering. Having first noted an association between sugar consumption and obesity, for instance, we could then go and study sugar consumption in a range of different places or among particular sub-groups. Interestingly, though, only one of Bradford Hill's criteria suggests that we perform an intervention to see what happens. Another suggests we can distinguish the cause from the effect by noting which occurs first, indicating his commitment to a temporal asymmetry of cause and effect.

Modern statistical approaches are considered more sophisticated than Bradford Hill's proposals. Statisticians believe that they can identify the cases of spurious correlation, such as the correlation between ice cream consumption and drowning. It can be found here that the correlation is not invariant under transformation of another factor, for instance warm weather. This is an application of the Causal Markov condition (Pearl 2000: 30). What we can find is that eating ice cream in cold weather does not lead to an increase in drowning incidents. And drowning in warm weather is more likely, but one is not more likely to drown if one eats an ice cream in either warm or cold weather. These data suggest, then, that warm weather is the significant factor. While it makes ice cream consumption more likely, it also makes swimming in pools and the sea more likely, which is what causally explains the raised incidence of drowning.

When scientists work to separate genuinely causal from accidental correlations, this is an acknowledgement, if only tacit, that causation is not the same as correlation. Hence, there is a distinction we can draw between 'correlation first' as a methodology and an ontological thesis that causation is nothing over and above correlation. The latter may draw some inspiration from Hume's view that the causal connection is unobservable as anything beyond regularity. By his empiricist strictures, this meant that it was illegitimate, because it was meaningless, to talk about causation as more than constant conjunction, with contiguity and temporal priority.

Such a view has some undoubted scientific appeal. Science deals with observable evidence, rather than with any unobservable metaphysical notions, so a scientist could well say that there is nothing more to causation than the evidence of correlation, as far as they are concerned. In contrast to Bradford Hill's informal use of criteria, then, one might argue that if one has the right kind of statistical data, it is enough to make a causal claim: a sufficiently robust correlation that holds in different circumstances and under different interventions.

The strongest form of such a view would be to say that this simply is causation, following Hume's regularity view. A compromise would acknowledge that causation is not identical with correlation but that if one had the right statistical techniques, one would at least find cases that are extensionally equivalent with real causal connections. This means that all and only those cases that are genuinely causal are so judged according to the right way of identifying correlations. Were this achieved, one might think that a large part of the scientific method was definitively resolved.

4.5 Collecting Evidence

Correlation data are crucial in epidemiology and other disciplines using statistical methods, such as econometrics. Okun's Law, for instance, tells us that with every 1 per cent fall in employment, a nation's GDP will fall 2 per cent below its potential. The example gives us one kind of case. This is where we apply our statistical analysis to observation data. The other kind of case is where we already have an idea of the causal connections we are looking for and run an experiment, for instance a randomized controlled trial (RCT). Here, we introduce a change and look to see if there is a recordable effect. We do this in order to confirm or deny a hypothesis.

The first kind of approach might be considered the less reliable but there are reasons why in some instances we cannot implement an intervention, either because it is impractical or unethical. Okun's law is derived from observation data alone, for example, because economies are too complicated for us to perform 'surgical' interventions, changing one factor alone (see Woodward 2003). In another example, it would be unethical to send people to the Elephant's Foot in Chernobyl to see whether they die. We can instead only record the histories of those who visited for other purposes. Where one cannot intervene, such as by running an RCT, one can still, with a sufficiently rich data set, run a regression analysis that aims to achieve much the same but does so through data analysis alone.

Either way, the idea is that in some circumstances we can confirm whether the predicted or anticipated effect is present, and then we have good reasons to say that the intervention worked. Causation is in some sense 'established'. In the opposite situation, one might conclude that there was no evidence found of the anticipated effect.

To say that this in either case proves that there is, or is not, causation is problematic. According to Popper (1959: 248), all we can ever do is corroborate a theory, where corroboration means that a theory has been tested and not falsified. But one also cannot conclude immediately from lack of evidence of causation that there is evidence of lack of causation. A more sophisticated approach needs to be taken to this latter question. One could try to judge how likely it would have been that, had A and B been causally connected, there would have been confirmation of it in a correlation. The more likely it is that there would be positive evidence, were the variables causally connected, then the more confident we should be of having falsified a causal hypothesis, if that evidence fails to be found.

What we have tried to show in this account of correlation is that Hume's regularity theory is very much alive within certain scientific approaches to finding causes. The constant conjunction view or regularity theory might be thought of as philosophically abandoned, in that few philosophers claim to hold it (Psillos 2002 being a possible exception). But the idea has not gone away and, as we have shown, can be found slightly modified in the idea that the search for causes should begin with correlation data. Regularity, in a different disguise, is correlation.

Five of the phenomena we identified as correlations can be understood as kinds of regularity. One was simple regularity itself, where A is always followed by B. The sciences offer something more sophisticated than that, but they are still explicable as regularities. Co-variance is a kind of regularity where one variable increases or decreases with another. One might add that it does so 'regularly' or 'whenever' there is an increase or decrease. Similarly, stable proportion is regularity in the sense that it remains unchanging through a number of different studies. Invariance, too, means a regular relationship repeated through a number of different values in the variables. Lastly, a constant is a kind of regularity: allegedly the most absolute, so regular, of all.

Although the regularity theory is easy to dismiss as being naïve—perhaps a theory that no one holds anymore—we can nevertheless see that various plausible features of it still persist in scientific methods based around the idea of causal laws being determined by the regular behaviour that we find in the world. It therefore remains worth exploring the philosophy behind the regularity theory of causation. We will continue to do so in the following three chapters.

5

Same Cause, Same Effect

5.1 Perfect Regularities

Correlation is clearly not the same as causation and we have just seen how causation cannot simply be derived from correlation data. Variables A and B might be correlated in some sense, but whether they are causally related and, if so, how, must be treated as separate questions. Nevertheless, we seem to think that correlations are a reliable symptom of causation, and that it is therefore useful to look for them when searching for causes. Common for all the types of correlations discussed in Chapter 4, is that they involve some form of *regularity*, which is also thought to be a central feature of causation: that a cause is something that brings about its effect in a regular manner. What, we now consider, is the source of this idea? We will follow it through a number of related notions that seem to underlie standard causal thinking.

Regularity seems a rational criterion for causation, and at least it is not something we would normally question. If the effect followed only occasionally and irregularly, then how would causal relations differ from accidental ones? Many correlations are known to be merely accidental. The one day that the bus is on time is the day you sleep in; carrying around a big umbrella keeps the rain away; a traffic jam occurs when you are running late; your friend arrives to the café just after you leave, and so on. Causal relations, we might think, ought to be more regular than this, and they should be more reliable. They should allow us to make predictions about the future based upon our knowledge about the present, for instance. This seems possible only if there is some degree of regularity between causes and effects.

When we think of causation in terms of regularity, it is because of two other features that we take to be essential for causation. The first is *robustness* and the second is *repeatability*. Robustness is contrasted with context-sensitivity and is the idea that a cause should be able to bring about its effect over a variety of contexts. A cause is thus something that pushes its effect through, across different situations. That we expect some degree of robustness is seen in some research methods, where we look at the same causal factor across a range of contexts to see if the same type of effect follows. For instance, one method for discovering the role of a gene is to check whether it produces a certain trait, or phenotype, with different background conditions. We might do this by placing it in various organisms, or by removing it, and then seeing what happens.

Repeatability is a requirement for robustness, since we could not test for it if a cause produced its effect only once. The idea of repeatability is explicitly expressed in Hume's regularity theory of causation, in which he linked causation to perfect correlations, or what he called the 'constant conjunction' of cause and effect. For Hume, there could therefore be no evidence of causation for a single and unique instance (Hume 1748: 148) because 'There must be a constant union betwixt the cause and effect. 'Tis chiefly this quality, that constitutes the relation' (Hume 1739: 173).

The expectation that causation involves regularity can thus be traced back to two related assumptions: that the effect must follow the cause several times (repeatability) and over a variety of situations or contexts (robustness). In this chapter, we will see how these and other features of Hume's regularity theory are deeply imbedded in the way we deal with causation in science.

5.2 The Folk Notion of Causation

If we consider these features—regularity, robustness, and repeatability—they seem motivated by an even more general principle or expectation that *the same cause should give the same effect*: a principle that we will now consider. Where does this expectation come from? Does it have an empirical basis? Or is it a philosophical assumption that this principle is part of the notion of causation?

Hume denies that there is anything in the world that makes causal regularity happen, at least we have no empirical grounds for assuming that there is. Still, he notes that there is a common habit or inclination in us to think that there is a necessary connection between cause and effect. When we observe a regularity, we expect that the same will happen in the future and that something in the world makes it so: some power, energy, or force (Hume 1739: 157). But this expectation is not backed up empirically, so Hume dismissed it as a metaphysical speculation (Hume 1739: 166). Or as Collins et al. (2004: 18) put it: that would be 'simply to confuse *guaranteeing* an outcome with *causing* that outcome'.

Following Russell's (1913) argument that causation is not an important notion for science, Norton (2007) is critical of what he calls 'the folk theory of causation', which typically involves the idea of necessity. 'The cause brings about the effect by necessity; this is expressed in the constancy of causation: the same causes always bring about the same effects' (p. 36). Kutach (2007) picks up on this point when he discusses causation and physical necessity.

The 'cement of the universe' is David Hume's famous phrase describing our conviction that causes necessitate their effects. Setting aside indeterminism momentarily, the empirical basis of this necessitation aspect of causation is captured by the principle 'same cause, same effect'. If we have two situations identical with respect to the precise cause and relevant environment, they both have the effect occurring.   (Kutach 2007: 329)

We see that three features are mentioned by Norton and Kutach as belonging to the folk notion of causation. One is that *causes necessitate their effects*, which Hume denied on empirical grounds. The second is *determinism*, although it is a separate philosophical debate whether it is a feature of causation or entirely independent of it (see for instance Anscombe 1971, Suárez and San Pedro 2011). The third is that *same cause gives same effect*, something that does not require necessity or determinism, but still is consistent with both. Hume is the source also of this principle, which he took to be empirically supported. 'The same cause always produces the same effect, and the same effect never arises but from the same cause. This principle we derive from experience, and is the source of most of our philosophical reasoning' (Hume 1739: I, iii, 15, 173; see also I, iii, 8, 105). But from what we saw in Chapter 4, this principle of *same cause, same effect* seems to be *prima facie* false. Very few cases of causation, if any at all, exhibit this feature completely, so it is unlikely that it could be empirically justified. For most of the correlations, such as co-variation, stable proportion, and invariance, there are some instances in which the cause does not succeed in bringing about its effect. Nevertheless, we maintain, with Kutach, that 'same cause, same effect' is part of the folk notion of causation. First, this can be seen in how less-than-perfect-regularities in data are explained or even marginalized, to which we will turn next. Furthermore, a number of strategies are used for making our causal claims universal and without exception, something that will be discussed in Chapter 6.

5.3 Dealing with Irregularities

Faced with a set of empirical data we will rarely see them lined up perfectly. The question is then how we should deal scientifically with irregularities displayed by real data. Popper (1959) had one suggestion, namely that whenever presented with a counterexample we should reject the theory and consider it falsified. The justification for this is a simple principle of logic. Since scientific hypotheses are universal, about all instances within the relevant domain, they are not verifiable even in principle. No set of observations could ever suffice to confirm that the hypothesis holds universally. On the other hand, a single observation of a counterexample would be enough to falsify it because it would show that it is not universally true. The theory should then be replaced by a better one, which we should again try to falsify through ongoing rigorous testing. Kuhn (1962), however, noted that this is not normal practice in science.

As has repeatedly been emphasized before, no theory ever solves all the puzzles with which it is confronted at a given time; nor are the solutions already achieved often perfect. On the contrary, it is just the incompleteness and imperfection of the existing data-theory fit that, at any time, define many of the puzzles that characterize normal science. If any and every failure to fit were ground for theory rejection, all theories ought to be rejected at all times.   (Kuhn 1962: 146)

Kuhn noticed that, more often than considering a theory falsified, one would hold on to it and explain away counterexamples by use of ad hoc hypotheses. Popper would allow such hypotheses only insofar as they increase the degree of falsifiability of the original theory rather than diminish it (Popper 1959: 62).

Popper's falsificationist approach has also been criticized by other philosophers of science. Among them Quine (1951) and Duhem (1954), who argued that no theory can be tested or falsified in isolation from the hypotheses, tools, and methods on which it relies. A counterexample would never target a particular hypothesis, but only a theory as a whole, including the test itself. This is now known as the Duhem-Quine thesis. Others, including Kuhn (1962: xiii), have argued that scientific progress is prevented if we don't allow a scientific theory to be developed in greater detail. Sometimes, what looks like a counterexample is in fact valuable scientific information (see Chapter 26). We might therefore want to be more cautious in judging theories falsified than suggested by Popper.

For most data, however, we are not dealing with a single counterexample, but many. Does this mean that we have sufficient grounds for discarding our theories because of a weight of evidence?

Consider, for instance, the difference between the data and the model in Figure 5.1 (from Henrique et al. 2009: 1583). While the model is perfectly linear, the data are scattered around it. The hypothesis is that the increase in protein proportion in body mass gain (dPRO/BMG) caused a reduction at the ratio between retained energy and metabolizable energy intake (RE/MEI). This is represented by the line, as a perfectly linear relationship, but the data show that there is only a 0.42 correlation. Had there been a complete overlap between the data and the model, the correlation would be 1.

There are a number of strategies for dealing with irregular data of this kind. We will here mention only some but there could be others.

5.3.1 Exceptions

It may be possible to dismiss as an exception a case that does not fit what is otherwise a general rule. However, this depends on what kind of rule it is. True identity and

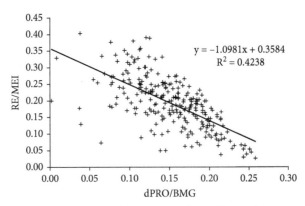

Figure 5.1 Model versus data. The y values represent how efficient the cow is in using energy to maintain and support the body processes (metabolism) and the x values represent the amount of protein over the total weight gain

classificatory rules do not admit exceptions but many others that are used in the natural sciences do.

Some equations in physics, for instance, do seem to give us perfect regularities. But an equation typically states a form of identity relation, for instance that $E = mc^2$ or that salt is sodium chloride. If there is identity, there could not be one without the other. Any case of salt must also be sodium chloride and there are, thus, no exceptions. That humans are mortal is usually regarded as a classificatory claim, meaning that anything that is not mortal will, for that very reason, not be human. It follows that there can be no exception.

However, another class of statements do seem to have generality while admitting the possibility of exceptions. As well as the 'is/are' of identity, there is also the 'is/are' of predication. We could for instance maintain that ravens are black while also allowing that some non-black ravens might exist, such as albinos (see Lowe 1982 and Drewery 2000). Such exceptions could be explicable by other factors. For instance, it might be possible to uphold the original theory, that ravens are black, but with a clause that the theory excludes a particular sub-class of ravens. It is notable that this class of statements, in which the possibility of exceptions is admitted, is characteristic of theories within the causal sciences. It is non-identity, non-classificatory claims that will be the subject of our theories.

5.3.2 Outliers

In statistics, there is a certain class of exceptions, called outliers. An outlier is an element of a data set that distinctly stands out from the rest of the data. This is illustrated in Figure 5.2 as the point at the bottom left. An outlier cannot be reconciled with, or predicted by, the model. When faced with an outlier, there is the question of how to regard it. Should it be included or excluded from the data? This is an unsettled debate in statistics (see for instance Cousineau and Chartier 2015, Altman and Krzywinski 2016), but there are clear arguments for why one might want to exclude them. If included, they will interfere with the statistical mean, standard deviation, and correlations, which could involve a misrepresentation of the situation. Say that we are looking at the average age of a consumer group of a new toy. If most of them are between 3 and 6, but one is 93, then the 93 year old could make the mean artificially high and certainly distorts the range. F. J. Anscombe (1968) calls this source of outliers 'multiple populations', which means that the outlier belongs to a highly deviant sub-group.

Outliers could also reveal a mistake in the data, in which case it should be corrected or removed. It could be a simple printing error, that the 93 year old was in fact 3. In Osborne and Overbay (2004), five types of errors are mentioned as sources of outliers: data error, misreporting, sampling error, standardization failure, and incorrect distribution assumption. But an outlier could also be a legitimate piece of data, although the chance of this is estimated to be 1 per cent when outlier

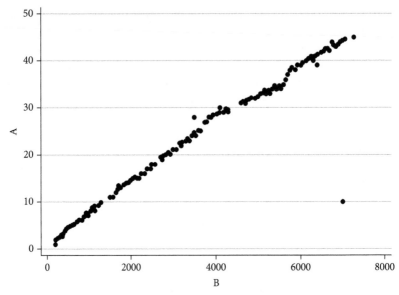

Figure 5.2 An outlier

is defined as three or more standard deviations from the statistical mean (Kenny 1979: 17). Legitimate outliers can be included, but perhaps not in the same model.

It might be considered a standard response to treat outliers as due to an error, since it is so unlikely that they were generated by the same process as the other data. To prevent an outlier from interfering with the results, one might then exclude it from the model. This would suggest that the outlier does not represent a genuine counterexample to the theory.

5.3.3 Noise or error

Almost all empirical studies will generate some data that cannot be predicted by the model. An outlier is one such example. But sometimes one will have a data set where the deviations from the model are too small to count as outliers. This is called noise or disturbance. In Figure 5.1, this phenomenon is represented by all the dots that don't line up perfectly with the model. The less noise, the closer we are to a perfect correlation, which is what the model represents. A model with no noise would give a correlation of 1.

Unlike an outlier, noise is to be expected and does not suggest that there is something wrong with the data. Instead, noise refers to the lack of perfect predictability of the model, which can be represented by the ϵ in the equation $y = f(x) + \epsilon$. The letter ϵ stands for *error* and refers to all the data that fall outside the model. In Chapter 6, we will see that a common strategy for dealing with noise, or error, is to add a *ceteris paribus* clause to our hypotheses. This way, we can assume some unspecified ideal conditions under which the model would hold universally.

5.3.4 Non-responders

In medical science there is a common phenomenon that patients don't respond to treatments in the expected way. These are referred to as non-responders. Although a treatment—usually a medication or a vaccine—is known to work, not everyone who gets it will experience an effect. Studies show that one of the most used drugs in the US, Nexium, taken for heartburn, only improves the condition for 1 in 25 patients, and Crestor, which lowers cholesterol, has an effect in 1 in 20 patients (Schork 2015). One reason for this is genetic variations, where a medication is developed to interact with a certain type of protein that the patient lacks. Other relevant factors are obesity, which can prevent the drug from having an effect, or even high metabolism, which might require a higher dosage. Gender also plays an important role, because men and women have different hormones, digestion, and body fat, that all affect the way in which a drug interacts with the body. Sometimes the variation in response has to do with heterogeneity of the disease itself, meaning that the disease has a different expression in different patients (Abelson and McLaughlin 2011).

All these variations in response have been used as an argument for personalized medicine, where a drug is tailored to the individual rather than to the group average. Still, the fact that there are non-responders is not interpreted as evidence that a treatment is ineffective; that is, as counterexamples to the theory. Rather, it is expected that all treatments will have some non-responders.

5.3.5 Interferers

One final approach for dealing with irregular data needs to be mentioned, which is to treat them as cases of interference. An interferer is a contextual factor that disturbs, alters, modifies, or disrupts the effect. In fact, some of the known reasons why patients might not get an effect of a medication, as discussed above, might best be treated as interferers. Hormones, stress, or fat can interfere with a drug, making it less effective. Interferers are also well known in experiments, especially the ones that are usually performed in the lab. The point of the lab is to control the contextual interferers that might affect or alter the results.

Interferers need not be considered as counterexamples to a theory, at least not if we can identify them and explain how they work against the effect in question. Many experiments are performed, not for proving a theory, but for learning purposes, such as in schools. It is then not uncommon that some of them fail to produce the predicted effect. Instead of interpreting the results as a disproof of chemistry or classical physics, we might identify a range of interferers, such as dust, dirt, friction, or unstable temperatures. Interferers should ideally be prevented from causally affecting a lab experiment.

What do these strategies tell us about our scientific ideals? One thing they show is that most scientists don't follow Popper's principle of strict falsification. But they also reveal a tacit commitment to the principle that same cause should ultimately give same effect in the scientific understanding of causation. Whenever one feels the need to explain away, abstract from, remove, or somehow deal with data that don't fit the general model, one seems to assume that the perfect model should be able to account for all the data. Any failure to make an accurate model would then be due to lack of detailed knowledge, or to a genuinely chancy element. If we knew how every factor affected the outcome, we would also be able to make a model that could predict the effect for each instance. Or, at least, this seems to be the tacit expectation.

5.4 Different Effect, Different Cause?

If we start with the assumption that same cause should give same effect, this would make us less inclined to accept counterexamples as genuine threats to a theory. Sometimes we have a high degree of confidence in a causal hypothesis, perhaps because the underlying mechanisms or principles are well known and understood. We know, for instance, why and how hitting a wineglass with a hammer is going to break it, and why and how pouring boiling oil over someone will burn and damage their skin. How would we respond, then, if the glass didn't break from the hammer or the skin remained unharmed by the oil? It seems that the more confident we are about a theory, the more evidence it would take before we considered it falsified, if we would at all. Counterexamples could always be explained away or marginalized, and we have already seen examples of such strategies.

If same cause gives same effect, then, by the logical rule of modus tollens, so is the following principle: *a difference in effect must mean that there was a difference in the cause.* Hume stated it like this:

The difference in the effects of two resembling objects must proceed from that particular, in which they differ. For as like causes always produce like effects, when in any instance we find out expectation to be disappointed, we must conclude that this irregularity proceeds from some difference in the causes.   (Hume 1739: I, iii, 15, 174)

Against Hume, we take this to be a metaphysical rather than an empirical assumption, since it is difficult to see how one could possibly reject it empirically. If causation happens in one case, but not in another, we typically take this to mean that something was different in those two causal set-ups. Otherwise, we should get the same outcome. This is related to an assumption of determinism, that the conditions of the present allow only one possible future (see Chapter 17). The only acceptable responses to a difference in the effect would then be that either (i) there was a difference in the causal set-up, or (ii) there was an element of chance involved.

Say we observe that two people smoke the same number of cigarettes throughout their lives. One gets cancer but the other does not. We could infer that one got lucky and the other got unlucky. But since determinism is often accepted as a prerequisite for causation and also for prediction, the most common response would be a variant on (i): that there is a causally relevant difference between the two smokers.

This type of reasoning is used in many empirical sciences, such as medicine and biology, where homogeneity is either assumed theoretically or produced in the lab. The aim is to have causal set-ups that remain the same. Lab rats should be genetically identical, for instance. Still, it is not always the case that they show identical responses. A behavioural study on lab mice showed that the same experiment with exactly the same inbred strain of mice in three different labs gave different results. Usually such differences could be explained away by a variation in tests, protocols, or unique features of the lab. What was special about this study, however, was that the researchers had made an effort to create identical settings simultaneously in the three labs. The researchers report: 'despite our efforts to equate laboratory environments, significant and, in some cases, large effects of site were found for nearly all variables' (Crabbe et al. 1999: 1670). They then go on to discuss which factors might have influenced the difference in behaviours between the labs, such as the experiments being performed by different people.

While this is a perfectly reasonable response to the variation in results, it shows how deeply embedded the principle of *same cause, same effect* is. The expectation remains that a perfectly detailed description of the causal set-ups would allow us to detect a difference in the cause whenever the effect differed. In Chapter 6, we will look at how this affects the way we think of causation as a basis for universal claims.

6

Under Ideal Conditions

6.1 Scientific Knowledge Is Universal and Abstract

Plato argued that the highest form of scientific knowledge—what he called *episteme* (Plato *Republic*: V, 533e–534a)—is universal and abstract rather than particular or material. The idea is still a cornerstone of science: the higher the degree of generalizability of our theory, the more valuable is the knowledge. The scientific interest we seem to have in particular facts or events is what we hope to be able to abstract or generalize from them. It might for instance be of some scientific interest to know the distance between the sun and each of the planets. But with Newton's three laws of motion and the law of universal gravitation, he could explain and predict all planetary motion in relation to the sun. By finding the underlying principles of motion, both in the heavens and on Earth, Newton set the standard for scientific knowledge for all his successors.

One reason we are interested in finding causes is the role they play in scientific laws. If we can uncover universal patterns of behaviour or laws of nature, then these could help explain and predict what may seem messy and accidental on the surface. The aim is then to find some fundamental principles that can unify a range of different phenomena. In this respect, science has been a success. The laws of Galileo and Newton purportedly allow us to predict with certainty what would happen, under specific conditions. The Newtonian forces permit a causal interpretation, and might even be used to argue that there was, in the end, some necessary connection between the two events involving Hume's billiard balls. Similar advances have also been made outside of physics. In biology, significant progress was made with Darwin's theory of evolution by natural selection, together with the discovery of DNA. And medicine would not be what it is without germ theory and the subsequent discovery of penicillin. Some laws have an explicit causal nature: heating iron bars causes them to expand; boiling water causes it to turn to steam; salt causes ice to melt. These are laws insofar as they aren't about individual objects or events, but allegedly apply to all instances at all times and places. What these few examples show is that general causal knowledge has been a game changer for science. Understanding causes is enlightening and empowering, and the more generally they apply, the more they enlighten and empower us.

6.2 The Messy Reality

If we think the best laws for our scientific purposes are those that are universally applicable and allow no exceptions, then theoretical physics seems to be in a privileged position. However, there is at least one plausible explanation of this. The laws of fundamental physics do seem exceptionless. But while reality is messy and admits a degree of irregularity, as outlined in Chapter 5, the laws of physics deal largely with ideal conditions. Whether the laws hold true for frictionless falls, perfect vacuums, or movements close to the speed of light, this is clearly not the world as we know it. A ball and a feather don't normally fall with the same speed, for instance. Only by stipulating a closed system with no friction, wind resistance, or presence of other interferers, will the laws of physics guarantee that the effect invariably follows. Our world, on the other hand, is full of friction and interferers. Rarely do we find a situation of perfect vacuum. On the contrary, it takes a lot of effort to produce it (see Close 2009: ch. 1). Likewise with motion close to the speed of light. Not many things travel with that speed, at least not naturally. Still, thanks to science, we now have reliable and universal knowledge about what would happen if anything ever were to move in a vacuum close to the speed of light.

Plato thought that the world of Forms was the most important, where all the abstract, universal truths can be found. But the world of Forms is not that of our empirical experience and one of the most damaging challenges to Platonism is the problem of how the ideal Forms relate to our concrete reality. This same challenge arises again when we consider how idealized laws relate to the messy reality. Once we move away from theoretical physics, we struggle to find neat, exceptionless laws. In the so-called special sciences, such as medicine, biology, psychology, sociology, anthropology, and economics, ideal conditions are hard to come across. This is perhaps expected, since things tend to get complex and messy when dealing with living nature, organisms, agents, and societies. In Chapter 5, we saw how most empirical data reveal a less than perfect regularity of what we take to be causes and effects. Instead, one can try to find some trends in the data by disregarding or bracketing off exceptions, outliers, interferers, non-respondents, and background noise. A motivation for this is that these trends, although they are not perfect laws, can nevertheless help us find some general, causal principles.

Studies show that irregular sleep patterns among students have a negative effect on learning (e.g. Hershner and Chervin 2014), and that the amount of charity donations is affected by available information about the receiver (Bachke et al. 2016). But even if we think that we can explain such trends causally, they are not suitable for making reliable predictions about individual students or individual donors. Is there a way in which one could make strict laws also outside fundamental physics? At least, we might expect that this should be possible in the natural sciences, since there is a closer relationship between physics, on the one hand, and chemistry, biochemistry, and molecular biology, on the other, than there is between physics and for instance economics or pedagogy.

6.3 Getting Laws from Irregularities

It should not surprise us that scientists struggle to find laws in the social sciences or in psychology. After all, we know that people are different, and perhaps even more so across cultures and societies than within. But according to philosophers such as Cartwright (1983, 1999) and Dupré (1993, 2001), reality, even the scientific one, is always messy. Any order is artificial, or even false. 'As Nancy Cartwright...has argued, simple physical laws, such as the law of universal gravitation, are literally false. Many, perhaps all, actual objects are subject to forces other than gravitation, and therefore do not behave in accordance with the laws of gravitation' (Dupré 2001: 12). The attempt to find laws that hold universally and without exceptions then seems a wasted effort. But is it inevitably so? As we saw in Chapter 5, there could be moves that explain away the exceptions and interferers. There are various strategies to help us retain the ideal of perfect laws, in practice or in theory.

6.3.1 Probabilistic laws

One strategy is to understand the laws as probabilistic (Woodward 2002: 304), and some philosophers have argued that causation itself is probabilistic (e.g. Reichenbach 1956, Suppes 1970, Mellor 1995). Looking at the available data, we could perhaps find a stable proportion of the effect given the cause, rather than a complete and perfect conjunction of cause and effect. And through comparative studies, we might find that an intervention raises the probability of the effect. Say, for instance, that a treatment cures 30 per cent of the patients. Instead of treating the remaining 70 per cent as counterexamples, one could say that this is a typical case of probabilistic causation. Such numbers could then be used for a generalization, with the advantage that we are back in business with our accurate predictions. Saying that 'C causes E with probability P' might therefore be an excellent candidate for a law that is universal and free from exceptions.

A disadvantage with probabilistic laws, however, is that they are less suitable for making predictions in individual cases. One could of course say that each individual will have a certain probability of an effect, but such predictions seem to become vacuously true. A doctor might tell a patient that she has an 8 per cent chance of a heart attack within the next four years. This would be unfalsifiable as a prediction, as noted by Popper (1959: 133). Whether or not she suffers a heart attack, it is not possible to say that anything was proven false. The problem persists even if a prediction specifies that the odds of the effect are one in a thousand. One might still be faced with that one exceptional case.

6.3.2 Nomological machines

A more practical strategy for making exceptionless claims is to adjust the way in which we perform scientific research. Since causes often fail to produce their effect because of the presence of interferers, we could try to create ideal and repeatable conditions. This could be done by isolating the cause from its known interferers, which is common in science. In the lab, for instance, with its sterile environment and genetically identical test animals or organisms, much effort is made to exclude interfering factors from the experiment. Cartwright (1999: ch. 3) calls such isolated set-ups 'nomological machines', since they are created with the goal of producing law-like behaviour. The way to do this is by producing—physically or theoretically— a closed system where all contextual factors are kept to a minimum. Cartwright explains it as follows: 'What is a nomological machine? It is a fixed (enough) arrangement of components, or factors, with stable (enough) capacities that in the right sort of stable (enough) environment will, with repeated operation, give rise to the kind of regular behaviour that we represent in our scientific laws' (Cartwright 1999: 50). Labs and other types of nomological machines are not restricted to the natural sciences. In experimental economics, one can use labs to perform controlled experiments and money games, although these are of a very different type than one would use in chemistry or biology. Still, the goal is the same. If the exact conditions in the lab were repeated, then we can predict with certainty that the same effect would follow. This is why it is important for us to try and keep the causal set-up stable and closed off from interfering elements.

A problem with doing all one's research in a lab is the inductive leap from predictions about the lab set-up to predictions about what will happen in the complexity of reality. What happens in the isolated lab and the controlled environment is usually very different from what happens in the field. This is natural, since the field will be an open system full of interferers, while the whole point of the lab is to exclude or shield off contextual disturbances that might interfere with and affect the outcome. As a result of this divergence, lab experiments often fail to reproduce outside the lab (see Chapter 28). As Cartwright (2007a) notes, there is a problem when the context in which we hunt for causes differs so much from the context in which we have to use them.

6.3.3 Ideal conditions

A third strategy for making universal laws is to simply specify some ideal conditions under which the cause would produce the effect without exceptions. How can we find such ideal conditions? One reasonable strategy could be by looking closer at all the successful outcomes where the cause actually did produce the effect, and then check whether they all have something in common. If we assume that same cause always gives same effect, it is natural to expect that there are at least some such conditions where the effect would follow. After all, there ought to be *some* causally relevant sense in which the successful cases are identical if the same cause was produced, even if they are different in other respects. Similarly, we might assume that all the unsuccessful cases also have something in common, such as an interferer.

A simple example is pregnancy. We know of many factors that need to be present for conception to happen, many of which are biological. But we also know that there are lots of lifestyle factors that can counteract the possibility of pregnancy, such as smoking, alcohol use, stress, eating disorders, being underweight, being overweight, or overexercise. So one way to explain why someone cannot get pregnant is by referring to a lack of one or more of the contributing factors, or the presence of one or more of the interfering factors. Instead of a law simply saying that C causes E, we could then specify it further to include the conditions under which it holds true, e.g. 'C causes E under $cond^1 \ldots cond^n$, excluding $interf^1 \ldots interf^n$'. This might, however, end up being a very long list of conditions. There might also be some yet unknown or unconfirmed factors that should have been included. We should therefore be careful to specify a complete list of conditions for C to guarantee E. A fourth strategy could then be introduced.

6.4 All Else Being Equal

If we are serious in making an exhaustive list of conditions under which a cause produces its effect without exceptions, it could end up being very long. Perhaps it would not even be possible to make a finite list. Possible interfering factors might include everything from a dust particle or drop of water to the implosion of the universe. For practical reasons, therefore, one might instead prefer to use a catch-all phrase for the list of ideal conditions. One possibility is then to add a clause to the law: 'C causes E, *ceteris paribus*', meaning 'all else being equal'. To use this clause explicitly is perhaps not so common in science, but an implicit assumption of such conditions is apparent in our research practices.

As discussed in Chapter 5, we quite generously allow counterexamples to causal claims without them falsifying our scientific hypotheses. Before concluding that a failed chemistry experiment represents a falsification of the hypothesis, we would for instance try to find out what went wrong. This seems rational, since we know that there are a number of possible interferers. School experiments sometimes fail to produce the expected outcome, and one might reason that this is because of a lack of both precision and sterile conditions. And in economics, one might stipulate that the causal laws only hold for rational agents, which is there taken to mean someone who aims to maximize their own profit. If someone has more altruistic motivations, this must be treated separately and explicitly, or instead be excluded from the model. By using statistical models, one can also stipulate ideal conditions such as normal distributions and averages. If we include enough of these generalizations and restrictions, we might be able to construe a set of conditions that makes the social sciences a bit more like theoretical physics.

A *ceteris paribus* clause is useful for this. It allows us to refer to an exemplary case where C successfully produced E and the conditions under which it did so, and then state that had these conditions been repeated, C would indeed produce E. Our prediction that E will follow C is then certain, as long as the conditions remain 'the same' or at least 'equal in the relevant sense'. The meaning and use of the *ceteris paribus* clause is a matter of philosophical controversy (Reutlinger et al. 2015), and it is not clear whether it should refer to 'ideal', 'same', or 'normal' conditions, or perhaps instead to 'a large set of' conditions that together with the cause makes it sufficient for the effect (Woodward 2002: 309). There are problems with each of these alternatives, and we will now turn to a couple of them.

6.5 Some Problems

We saw that causal laws are valuable to us because they are general, or even universal. But if we want to claim laws as universally applicable, at least the conditions under which they hold must be repeatable. Suppose, therefore, that we think that C causes E means the same as that C *normally* succeeds in producing E. This could then be one meaning of the *ceteris paribus* clause, but perhaps not one that is suitable for physics. If we look at theoretical physics, there is not much that is normal about the conditions under which the laws hold. On the contrary, they are sometimes practically impossible to produce. Many experiments in physics are therefore thought experiments, or the laws are derived from results of experiments in less than ideal conditions.

If we instead use the *ceteris paribus* clause to refer to *ideal* conditions, this also becomes difficult, at least once we go outside the realm of theoretical physics. What are the ideal conditions for learning to read, for instance? Or for becoming pregnant? Many effects seem to be produced under less than ideal conditions, while sometimes even ideal conditions are insufficient to produce an effect. Matches are sometimes lit in wind and rain, some women get pregnant even when on birth control, and some children go to school in war zones and still learn to read and write. On the other hand, matches sometimes don't light even indoors, some women don't get pregnant even with fertility treatment, and many kids struggle to concentrate on learning even in safe, quiet, and academically stimulating environments. Calling the first set of conditions 'ideal' could only mean that they were sufficient in the circumstances to produce the effect, and ideal only in that respect. Ideal conditions would then have to refer to the conditions under which E actually follows from C, not some common set of conditions that hold for all instances. In Cartwright's explication of the nomological machine, in the quote above, we saw that the level of isolation and control only has to be fixed or stable 'enough'. What counts as enough will typically depend on whether or not the predicted outcome follows. This makes the so-called ideal conditions entirely relative to the production of the effect, as an implicit success criterion of causation. Causal predictions then become vacuously true.

Again, this all comes down to the idea that the same cause should always give the same effect, either because we think causation is about perfect regularities, or because we believe that causes somehow necessitate their effects. To say that a cause was 'sufficient in the circumstances' for the effect, is one way of saying that causes necessitate or guarantee their effects, at least under some conditions (we challenge causal necessitation in Part III of this book). This idea was proposed most famously by Mackie (1980: 62), defining a cause as 'an *insufficient* but *non-redundant* part of an *unnecessary* but *sufficient* condition' for the effect. He called this an INUS condition. It is thus an important difference between a necessary condition and a sufficient condition. A necessary condition is a *sine qua non*, meaning 'without which, not'. This is what is referred to when we speak of causation as counterfactual dependence: if it weren't for C, E would not have happened. These are the types of causes we find when we use randomized controlled trials, where we compare the result of a trial intervention with an absence of that intervention. Mackie's INUS conditions are sufficient causes, which means that they are themselves enough to produce the effect. Here, this view is more sophisticated, since the cause is one of many other factors that together are sufficient for the effect. This means that the *ceteris paribus* clause would refer to the set of conditions that together with the cause guarantees the production of the effect. The set of conditions and the cause would not be able to produce the effect on their own, but only jointly. But as long as the conditions are right, the effect is guaranteed.

It seems, then, that also Mackie's INUS conditions come with a *ceteris paribus* clause, but only implicitly. If E does not follow from C, we have to assume that the conditions were not right. Otherwise, it should be guaranteed. Again, any such causal claim risks becoming irrefutable in principle. The *ceteris paribus* clause that makes our causal claims law-like, and without exceptions, thus seems to remove what we wanted to achieve by using it: reliable and certain predictions. If we claim that the effect is *guaranteed* to follow from the cause, but only whenever it actually does follow (since it often does not), then how are we better off than we were before?

7

One Effect, One Cause?

7.1 Monocausality

A social scientist might wonder what caused the global economic crisis of 2007–8. Was it selfish bankers, too much borrowing from individual consumers, a lack of financial regulation, an inevitable outcome of a socioeconomic system containing inherent contradictions, or something else? Perhaps a sensible answer would be to say that the crisis was caused by many of these factors all working together. Nevertheless, we often look for *the* cause—singular—which at least suggests the idea that for each effect there is just one cause.

We will argue in this chapter that the matter is more complicated. In maybe all cases of causation, there are multiple causes at work. The idea of monocausality has, however, been enshrined in some standard ways of thinking about causation, including in the philosophy of causation. Even if that can be disregarded as a relatively superficial presentation of the matter—one that we tacitly accept to be an oversimplification—it is still a view that can skew our thinking in harmful ways.

There might be an idea, for instance, that we are best to understand a single case of causation first, abstracted away from all the accompanying background conditions, because that is causation in its purest form. And then, generalizing, we may have found that causation always issues in a regularity; or at least some say. Complexity might then be dismissed as a relatively unimportant detail. If we understand single cases of causation, isn't complexity just many instances of the same sort of thing, all happening at the same time? We argue that this approach will not do. If we try to understand causation even in one case, abstracted from its complex setting, then there is a danger that we already accept some controversial assumptions. We may treat complexity as just an addition or aggregate of a set of simple causes, and this might not be good enough to account for many standard cases. Furthermore, we need to consider the causal selection problem: an issue that is raised only once we encounter causal complexity, as we will show. Causal selection can even have a political dimension, where such normative considerations are aided by fostering a myth of monocausality.

It is easy to think of an effect as having one cause, especially as we are now more tempted to focus on the efficient cause alone. The notion of efficient cause comes from Aristotle (*Physics* II.3, 194b24ff) although he saw it as only one of four different types of cause. We can think of the efficient cause as the stimulating cause: it is the change that, in the circumstances, precipitated or brought about the effect. In almost every case, we will think of this as a single factor and the difference-maker. The other factors that were already in place can be classed as background conditions.

Some simple examples illustrate this notion. A match couldn't light but for the presence of oxygen and the flammable tip. But it is only the striking of the match against the side of the box that triggers it to light. The efficient cause of lighting is the striking of the match. Similarly, a book might have been placed overhanging the edge of a desk, precariously, but it is the nudging of it by someone's elbow, and only that, which causes it to fall. Sometimes this efficient cause is itself complicated and spread over time. Marie Curie could have had a genetic predisposition towards aplastic anaemia (leukaemia), for example, but we might still say that there was a trigger of the disease, such as the long-term exposure to ionizing radiation. The thinking here is that radiation exposure was the factor that started the onset of the disease whereas the genetic make-up would be background only. There is supposed to be, then, just one cause of the effect.

A number of our philosophical theories seem to reinforce this idea. For instance, on Lewis' (1973b) counterfactual dependence theory, causes are supposed to make a difference. The difference is that the effect would not have happened without the cause. This creates a problem in cases where a second cause is present that could also have produced the effect. A manipulationist theory of causation (Woodward 2003) tells us that we should think of a cause as what would be an intervention that led to a change in the effect. And a transference theory of causation (Fair 1979, Salmon 1984, Dowe 2000, Kistler 2006) encourages us to think of there being a transfer, of some mark such as energy or momentum, that can be traced from the cause to the effect.

None of these accounts are committed to the idea that there is only one cause for each effect. Indeed, Lewis allows that an effect typically has many causes. The counterfactual dependence account does have an issue with some kinds of multiple causality, as we shall see in Chapter 15. We can certainly say, however, that none of these accounts foreground causal complexity within their theories. Rather, it is treated as matter that can be kept in the background without any serious detriment to the account. This itself is a mistake, we argue.

7.2 Multiple Causality

What, then, do we mean by causal complexity and what is this vital role that it should play in our understanding and theories?

First we will distinguish two kinds of multiple causality as we will see that their implications differ. We will call these two kinds joint causes and overdetermining causes.

Overdetermining causes will gain more attention (in Section 15.2) so we will not discuss them in too much detail here. These are cases where more than one cause is present that is able to produce the same effect even if the other overdetermining cause were absent. There can be more than two overdeterminers in some cases. A simple example of overdetermination is a bookshelf that is attached to a wall by brackets affixed with screws. The shelf can stay up even if one or two screws come lose and fall out; so, unless that happens, the shelf maintaining its position is overdetermined by the screws. What is characteristic of overdetermination is not just that there are multiple causes involved but that there is also some degree of causal redundancy. At least some part of the cause was not needed for the effect to have occurred. The shelf is affixed with twenty-four screws but sixteen might have been adequate, for example, for every load the shelf bears in its lifetime. Such redundancy can be grasped with the notion that there could have been less in the cause and the effect still occur. This part could have been removed. It thus added nothing to the effect, which would have occurred even without it.

This aspect distinguishes overdetermination from more standard examples of causal complexity, the cases of joint causes. Joint causation is when an effect is produced by a number of causes working together where none of them are redundant. The effect occurs, therefore, only because those causes worked together, the effect being their joint action. Suppose two people lift a sofa completely off the ground, one at each end. Neither of them could have lifted it alone, at least not in the same way. The action of both people was thus needed. Similarly, a forest fire could start both because someone dropped a match but also because the forest was dry and warm and thus apt to set alight. A match dropped in the rainy season might have had no significant effect. It looks, then, like one needs both and that the fire could be considered the joint manifestation of the dry foliage and the match.

The forest fire will also need a further causal factor. Wood can burn only in the presence of oxygen. Might we then think of oxygen as a third cause? And what, also, of the chemical composition of wood? Is that a cause, given that the trees would not have burnt had they been made of something else, such as metal? One could then think of the causes as being large in number, taking into account all the relevant properties of the match that was lit and dropped, the forest and all the background enabling conditions. Perhaps even the laws of nature should be included among the causes, given that history might be different if the laws of nature were changed or did not apply in this case.

Aristotle recognized that causes came in different categories. The dry wood of the trees should be considered a cause of the fire, because it matters what they are made of if they are to burn. But also the type of trees and the density of the forest play a causal role. Are the trees close enough together to allow flames to pass from one to the other? Do they allow the wind through to fan the flames? The material cause is the matter of which something is made and the formal cause its nature or arrangement: the form the wood takes, in this case. The efficient cause is something that is added to the material and formal cause; it is the change that precipitates the effect. Here, it was the dropping of a match. The final cause concerns the purpose, or *telos*, of causation; that towards which the cause is directed. Bringing together flame and wood matters only because, when they are together, they are directed toward burning. This is the end toward which flame and wood will dispose.

Since the advent of so called modern philosophy, Aristotelians, and the Thomists who follow Aristotle's philosophy, believe that we have concentrated overly on efficient cause alone, ignoring the true complexity involved in the production of an effect (Feser 2015). Causation has been effectively reduced to the last difference-maker that occurs before the effect. What, then, of all the other factors that must be present in order for the effect to occur? One common strategy that persists is to draw a distinction between causes and background conditions. One could urge that there is just one cause of an effect (except in overdetermination cases) and all the other factors that are needed for the effect are merely background conditions. This sort of distinction need not automatically rule out complex causes. One might be able to say that the one change that initiated a causal process was itself complex. Perhaps the effect is triggered only if a combination of circumstances happen to be in place. But, even if the efficient cause is a complex, it is still distinguished from all the other kinds of cause, which are relegated to background conditions.

7.3 The Causal Selection Problem

How we can do this remains a contentious issue. How does one select, from among many candidates, which factor is the cause of an effect and which is a mere background condition? Should one think of the efficient cause as the last contributor to an effect that is put in place immediately prior to the effect occurring? Causes are often referred to as triggers, for instance of migraine, epileptic seizures, eczema, climate change, market collapse, or political conflicts. On that basis, one might say that the prior factors are background conditions and the striking of the match, as the last thing in place, is the triggering cause.

A problem with such a view is that all factors have to be present at the start of the effect in order to exercise any influence upon it. Does it matter in which order they were assembled? Would it matter at all if oxygen was introduced at the moment that the match was struck, in which case the presence of oxygen has equal claim to being the cause as does the striking of the match? Might one just say, then, that the cause is the unusual or atypical occurrence? The presence of oxygen, in almost all circumstances, can be assumed. That's what makes it a background condition. A cause is, in some sense, a change from the ordinary or a deviation from the norm. A problem here, then, is how to account for cases of causation that occur in unusual circumstances. If a man is suffocating, for instance, then the introduction of oxygen can be what saves him. Is this any less of a cause just because the presence of oxygen is usually expected?

There are three ways in which the basis of causal selection can be understood. The first we might think of as rationalist or ontological. On this view there is a principled reason why one factor stands out and deserves to be thought of as a cause of an effect instead of merely a background condition for it. We have seen a few such proposals, such as that the cause was the one last added. In this category, we can also place Mackie's (1974: 62) INUS condition proposal; the thesis that a cause is an *Insufficient* but *Necessary* part of a condition that is itself *Unnecessary* but *Sufficient* for the effect. Suppose a building burns down and the fire investigator says it was due to an electrical fault. According to Mackie's analysis this means the following. It wasn't necessary that the fire be started by an electrical fault in that there are other ways in which buildings can burn down, such as a gas explosion or arson. However, in this case the electrical fault did start the fire, and was therefore sufficient for it. This is what is meant by an unnecessary but sufficient condition for the fire (the *US* clause). And we can also say that the electrical fault was an insufficient but necessary part of this particular, electrical fault-way of starting a fire (the *IN* clause). On Mackie's view, then, the distinction between cause and background condition has a rational basis, since background conditions would not meet the INUS condition for being a cause. We can also say that the distinction has an ontological basis in that it is facts about the world that determine whether something is a cause or not; it is more than just a matter of our own interests.

The second way, a contextualist view, would have it slightly differently. On such an account (discussed by Mackie 1974: 34, 35, Lewis 1986a: 162, and Schaffer 2012), what is identified as the cause of a particular effect is relative to the context in which a question is asked, which could include who is asking the question and of whom. Hence, if we ask how a footballer's leg was broken, a physiotherapist might say it was due to the player's foot being twisted outwards when his bodyweight was all on that leg. A referee might say that the break was caused by a reckless tackle by an opponent who slid in from the side. And a fan might say that the leg was broken in a desperate late attempt to stop a goal. These are three different candidate explanations and all could be valid and true within a certain context. However, this makes the question of what caused what at least partly an epistemological matter. The answer depends on what is known, believed, and expected by those investigating the aetiology of the explanans. These can still be rational matters, though, with good and bad, appropriate and inappropriate causal explanations within those contexts, but those reasons are essentially matters concerning our epistemic states.

A third way of picking out one cause among many candidates as *the* cause, can be thought of as non-factive. Normative causation can be classed this way. Consider the causes of poverty. Some say that it is due to unfair social structures or political decisions on wealth distribution. Others say it is due to individual failings whereby people become poor because they are lazy or untalented. People might favour one of these two options as the cause in different proportions in different groups. Research confirming this view goes back to Feagin (1972) but is ongoing, for example da Costa and Dias (2015). Perhaps a sociology student is more likely to blame social structures for poverty, but a conservative is more likely to blame individual failings. This raises the prospect that it is at least in part a moral or political decision what to identify as a cause of something; that is, a normative matter. Consider genetic research into the causes of cancer, which receives public and other sources of funding. From one point of view, this research programme tacitly locates the 'fault' for cancer in the individual sufferer. In a way, however, we already know perfectly well some of the main causes of cancer and could perform interventions that would produce a huge decrease in cancer rates, for example, if we outlaw the tobacco industry, and other known producers of carcinogens. Or perhaps we should blame the capitalist system itself (for why else other than financial profits would such industries and products by tolerated?). The selection of one factor as cause, at the expense of others that could also have a claim, can thus reflect the values or norms of those making the identification.[1]

We will not attempt to resolve which of the ontological, contextualist, or normative accounts is best. Indeed, all three make sense and provide a basis on which causes can fairly be distinguished from conditions. We have our own preferred ontological basis, for instance, that distinguishes causes from mere *sine qua non* conditions. There is a difference, we maintain, between a cause which disposes toward an effect and is capable of producing it, and something that is merely a necessary condition for an effect without having a disposition toward it. The Big Bang is an alleged necessary condition for everything that followed it, for instance, in that nothing at all would be here had it not happened. But it did not cause the creation of Obamacare because it contained no disposition toward it happening (nor one against it happening). We also accept that there can be good contextual and normative reasons for identifying something as a cause, however.

7.4 Complexity Accepted

Despite the foregoing account, which offers a degree of legitimacy to the singling out of a factor or factors as the cause of an effect, we want to end this chapter by making the case for the acceptance of complexity in causation.

One might think it harmless to start first with an account of what it is for one thing to be a cause of an effect. However, if we move to speak of *the* cause, then it can suggest that it is the *only* cause. This is problematic and it also feeds the idea of causation involving a perfect regularity. The idea of monocausality—each effect having just one cause—is problematic in that it cannot account for the context-sensitivity of causes. Something that in one context produces an effect can fail to do so in a different context: sometimes only a slightly different context. The taking of a pill in one context might alleviate pain but, in another, cause it, such as when someone gets stomach cramps from the painkiller ibuprofen. We will say more about such context-sensitivity, and its implications, in Part III.

---

[1] This idea has been developed in as yet unpublished work by Alex Kaiserman and Robin Zheng.

There is an even stronger reason why we should take account of complexity, however; indeed, why we should place it at the core of our theory. This is that the causal powers of wholes are not always simple aggregates of the causal powers of their parts. We cannot, then, just add together a number of single causal facts, understood in isolation, and get an understanding of the whole that they form, for there is something essentially complex and related about those wholes.

Consider a simple causal structure, such as an archway formed of stones. Each stone has its own cluster of causal powers. It is capable of breaking a window, for instance, though it is not capable of defying gravity. An individual stone in the arch is incapable of staying where it is without the position of all the other stones in the arch. If the keystone is removed, then the arch will come crashing down. This power of the whole arch, to form an upward curve that bridges a gap, with no support underneath, is a power that arises only when there are multiple components that stand in a particular arrangement.

This is a simple but not isolated example. Chemical composition provides another (see Mill 1843: III, xi). Chemical components do not simply aggregate or add. They undergo a transformation when they bond, changing each other. There are some other crucial cases like this, which arguably qualify as cases of emergent phenomena. There are various 'ingredients' that make up a living organism, for instance, none of which are particularly rare. What is truly exceptional is the very special arrangements in which those ingredients have to stand in order to constitute and sustain a living thing. This arrangement is so finely tuned that one very small difference can lead to a different kind of organism, or no living thing at all. It is the arrangement that is the main task for science to uncover, not only the ingredients. Similar things can be said about other emergent phenomena, such as how mental causal powers can emerge from the physical matter in brains, but only when they are in that very specific arrangement. We take this theme up again in Chapter 14.

Complexity in causation cannot be overlooked, therefore. One motivation for holism over reductionism is the idea that there are some things that can only be done by wholes whose parts are cooperating with a very specific form of dynamic organization. From the same set of parts and their causal powers, many differently empowered wholes could be formed. Some of those organizations can be regarded as special when we see new features emerging, not found among the parts individually. Those features show us that to think in terms of one cause for every effect will impose limits on our understanding.

PART III

Interference and Prevention

8

Have Your Cause and Beat It

8.1 Inevitability?

Suppose you were to contract a disease which many times in the past, when others have caught it, has been deadly. Should you resign yourself to your fate? Does it mean that there is nothing you can do? A routine physical process that we have the power to trigger, and which is otherwise useful to us, leads to the emission of potentially harmful radiation. Despite any potential benefits of this process, should we just ban it? Finally, an area on the side of a mountain suffers regular mudslides after wet weather. Do we just have to let it happen? It may seem strange to ask such questions but we hope their importance will soon become clear.

There are many instances where past experience is believed to show that something has always been the case and perhaps this leads us to believe that causation is in play. Deaths have been caused by diseases, mudslides have been caused by wet weather, and radiation is emitted as part of the process of x-ray photography used in medicine and for security screening. But does this mean that the effect in question must always be produced by the cause, once the cause is in place? Is the effect then inevitable?

There is a view that the world is to a degree regular and predictable. This may tempt us to think of regularity as the sign of causation, hence what we should be looking for in the scientific quest to uncover causal connections. A causal connection between two phenomena would explain the empirical regularity, which could then be taken as a reliable indicator of causation being at work. A primary job of science would then be to record the regularities found in the world and we would need methods for doing so: gathering data and looking for trends using statistical analysis, for instance. But is this view grounded in a credible philosophy of causation? Does it reflect what causation actually is? We hope to show that the truth is more complicated.

Perhaps there are not many who follow explicitly Hume's constant conjunction theory of cause. But we have already shown, in Part II, that the basic idea of regularity in causation persists: in notions such as correlation and in the idea that the real regularities can be found if we isolate causes or create ideal conditions. We also posed some problems for this view, however. The same cause does not always produce the same effect and there are irregularities to be found even when there is otherwise a statistical trend. We may, nevertheless, try to salvage the idea of regularity in causation by adding *ceteris paribus* qualifications to our causal claims. If only everything were set up correctly, then the effect would always follow the cause; so it is normal to think.

Part III will now take us in a different direction. To begin with, the aims of the current chapter are twofold. First, we urge that perfect regularity was never a worthy goal of a theory of causation. This is because it clashes with a fundamental tenet of causal understanding in the sciences, namely that it is possible to intervene and change the course of events. We will say more about this. Second, we aim to provide the theoretical background that explains why we don't get perfect regularities out of causes and, thus, why those who try to preserve the spirit of the regularity theory are inevitably obliged to speak in terms of ideal conditions, or some related notion, for instance when all else is equal or there is an expanded 'total' cause. We will explain how such qualifications attempt, unsuccessfully, to rescue the idea that causation issues in perfect regularities.

The argument of this chapter occurs within the context of Part III, as a whole, offering a tendential alternative to understanding the world in terms of perfect regularity and explaining exactly why context matters to causes.

8.2 The Power of Intervention

Predictions are notoriously fallible. Sometimes we have to predict the behaviour of very complicated subjects, e.g. how a person will respond in a certain social situation. But even when dealing with purely physical processes in the natural world, we know that they can be very complex and we cannot be entirely certain what will follow what. This could be considered a problem in that we do not have complete foresight over the course of events nor, in that case, complete control. If we attempt to control some particular causal processes, but cannot predict with entire accuracy what will follow, then any such control is clearly less than total. Hence, we see that the notion of control is dependent on predictability, and thus is reduced in proportion with unpredictability.

We can offer a reason why causes sometimes fail to produce their typical effects (we offer another reason in Chapter 10). Causation can be interfered with and sometimes prevented from occurring altogether. Hence, a typical cause could be in place and might even begin the process of producing its effect, but if that process is interfered with, or prevented, then the typical effect of that cause still need not occur.

We make use of a distinction here: prevention versus interference. We would explain it as follows. Prevention stops something from happening; interference allows something to happen but it happens in a different way, extent, or time. For example, iron, and alloys containing iron, tend to rust. But rusting can be prevented from happening, for example by galvanizing the alloy. If the galvanization works, then the alloy does not rust. Alternately, we might interfere with the rusting in a way that doesn't prevent it completely but slows it down; for example, if we add a corrosion inhibitor. We concede that this distinction between prevention and interference is not completely sharp. It may be a distinction that applies just on the basis of how detailed is the specification of the effect. In interference, one is effectively preventing a very precisely specified effect occurring even if a similar though slightly different one occurs. Hence, in a case we could think of as interference, some iron rusting after five years may be prevented from occurring, but rusting after ten years occurs in its place. So for many of the examples we will discuss here, we need not take the trouble to adhere rigidly to the distinction between prevention and interference.

The possibility of prevention can be a bad thing, we have seen, if it is important to us that the expected effect is produced. We always want aeroplanes to stay aloft through their forward and upward thrust and that an antibiotic will kill a bacterial infection. Anything that interferes with the desired causal process, such as antibiotic resistance, is thus a threat to these desired outcomes. But the opportunity of prevention can also be to our advantage, where the effect is something that we want to avoid. If we can prevent a disease killing someone, or rain causing a mudslide, then it is all to the good. Interference in the otherwise regular course of events then comes to be thought of as an intervention or manipulation, especially if human agents are the ones interfering.

The possibility of intervention can be crucial. If causes produced perfect regularities then we would be powerless to prevent them. Whenever there was C, there would be E. This is another way of saying that there would be the regularity of C always followed by E, which is what we mean by a perfect regularity. Of course, we could, in that case, try to prevent C from occurring in the first place, if E is something one needed to avoid. For example, suppose that if a nuclear reactor goes critical, the core will melt, releasing lethal doses of radiation. In that case, one's best approach is to prevent the reactor going critical.

But there are other cases where C may have occurred already or be unavoidable, so we cannot stop it. Or C might have otherwise desirable effects that we wish to exploit. What does one then do if the undesirable E always follows C? It seems one has to accept the inevitable doom of the disease, or the radiation of x-ray photography. That is at least what a pure regularity view seems to imply.

However, the possibility of prevention and interference can be our saviour from the inevitability of undesirable causation. Since causes typically interact and produce different outcomes in different contexts, it is possible for us to intervene to avoid an unwanted effect. We can also intervene to bring about the effect we do want. This is why there are philosophers who understand causation in terms of intervention or manipulability (for instance Woodward 2003).

Sometimes we need a whole set of advanced tools and instruments to perform the manipulation, such as those we assemble under laboratory conditions. A striking example is, as already mentioned, the vast mechanism and organization, spending a huge budget, that is required to collide hadrons with adequate control at CERN. The cheaper option is where some manipulations are simply simulated in a model for purposes of prediction, and not performed physically. One could change a variable in a computer simulation and see what subsequently changes with it. Even more distanced from reality are thought experiments, such as the twin experiment of relativity theory, where one twin is sent out into space, travelling at near light speed, while the other stays home. There are many reasons why we cannot currently perform this experiment in reality.

Much of the idea of technology depends on the idea of intervention. Natural causes can, for example, be prevented from having their otherwise natural effects if we have the right technology. We can put defences in place to hinder coastal erosion, fighting against nature. Or consider a technology that requires the counteracting of one effect by another. The main rotor of a helicopter exerts an upward thrust and lifts the vehicle off the ground. However, the rotor also exerts a torque force, which would spin the helicopter around in the opposite direction to the turning of the blades. But this effect can be prevented: the best-known method being a vertically angled smaller anti-torque rotor on the tail that pushed back against the main torque force, keeping the helicopter's orientation steady. Or, as in the case of x-ray photography, we can simply put up shields so that the machine operator does not have multiple exposures to low-level radiation.

We can see, then, why there is a close connection between the notions of cause and intervention. It could be explained like so: 'a very central part of the commonsense notion of cause is precisely that causes are potential handles or devices for bringing about effects' (Woodward 2003: 12). The experimental method exploits exactly this feature of causation. By manipulating the causal set-up, we can produce an effect that wouldn't happen otherwise; that is, without our intervention. Hence, we are able to intervene in a system and see what changes with that intervention. This provides one method of detecting what causes what, to which we will return in Chapter 24.

8.3 Subtractive and Additive Interference

There is a distinction that can be made between additive and subtractive interference, and similarly between additive and subtractive prevention. We will see that this distinction is important for understanding how causation works.

If we want to counteract an effect, whether it is in real-world practice or within the theoretical confines of a model, we can do so in two ways. The first way is to remove a causal factor from the complex of circumstances that we think of as the cause. This is, hence, a subtraction: a taking away from or removal of something from the cause. It is clear that this can be done for some cases of causation. Consider the mudslide case again. Here something can be subtracted from the situation that has been leading to the mudslides after rain, for instance, digging out the earth before it rains so that there is no subsequent slide. Similar in principle is the prevention of cancer by removing one of its causes, such as cigarette smoking.

But it is not always possible to interfere in this way. For practical reasons, some causes may be beyond our control to remove in the circumstances. We cannot turn off the radiation that is emitted from x-ray photography (radiographs): it is an essential part of the x-ray process that radiation is emitted. The radiograph image would not be taken without it. We have to accept, therefore, that there will be a radiation emission. An even simpler example is that we cannot turn off the sun that warms a room.

There is, however, another option. We can add an extra factor that tends away from the unwanted effect. In the case of a room being overheated by the sun, we can switch on an air-conditioner. The sun is still shining on the same spot, but cool air is added. Likewise, we cannot remove the torque force of the helicopter main rotor: it is a consequence of the conservation of angular momentum. So the strategy, correctly, is to introduce a counterbalancing force from a vertical rotor. Hence, this is an addition to the cause—the introduction of a new causal factor—as opposed to the subtractive interference discussed above.

We thus have two different kinds of interference: subtractive (removing the cause) and additive (adding something more to the cause), which can stop the effect from occurring (see also Section 9.2). In medicine, for example, both of these strategies can come into play. Additive and subtractive interventions are prescribed. A pacemaker can be added to a struggling heart, usually with some medication in support. But the doctor can also tell someone to quit drinking and smoking to remove some of the factors that put strain on the heart: hence, a subtractive interference. In molecular biology, genetically modified organisms are also produced by causal interference, by removing or adding genes, and thus deviating existing DNA from its otherwise natural development. Genes can also be 'switched off' or 'activated'.

Note how interferer and preventer can be understood as relative notions. We can see this with the counterbalancing forces of the helicopter rotor blades. Force A is prevented from having its full effect by force B, and force B is prevented from having its full effect by force A. They are thus each other's interferers. Sometimes, perfectly balanced interferers produce an equilibrium state of no net change, which is a variety of causation of absence (see Chapter 16).

We can see now that this distinction between subtractive and additive interference is crucial. What is significant about additive interference cases is that they allow you to have your cause and beat it; that is, although the cause is still present, because one doesn't want its typical effect, one adds something else as well that prevents the effect from occurring or changes it. Therefore, if we go back to the questions with which this chapter began, we can see that there is no need to accept the inevitability of an effect even in those cases where the cause is already present. If we have the cause, perhaps one that is impossible to remove, but then *add* an interferer, we might not get the effect after all. In science, there seem to be endless possibilities for additive interference, in practice or at least in theory. And if there aren't any actual preventers yet known, chances are that someone is looking to find them, if it is desirable to avoid a specific effect.

There might, nevertheless, still appear to be strict or perfect regularity in these and other types of case, despite the possibility of interference that we have outlined. This appearance could be for a variety of reasons:

1. Where no preventer of E, given C, is currently known, though future technology may yet identify one. For example, as of 2018 there is no known cure for HIV/AIDS, or even the common cold, but people are working to find one. They do so on the assumption that there is at least a possible cure and their task is to find it.

2. We might consider a cause so late in the process of it causing E that E then seems inevitable, because its probability approaches 1. For example: crossing the street involves low risk of causing death and there are various measures we can take to prevent it. However, if one steps in front of a moving vehicle, the chance of death is considerably increased. And if one is just 1 metre away from a car travelling at 60 km/h, death looks inevitable. But clearly it was preventable (more easily) earlier in the succession of events that led to the death.

3. We can mistake non-causal necessity for causal. There are some examples of necessity in nature, such as that water is necessarily H_2O, according to one account (Kripke 1980). But this necessity comes, we have argued (Mumford and Anjum 2011: 167), from the identity of water with H_2O; not from causation. H_2O does not cause water. Another example of non-causal necessities could be essences, for instance if we think that humans are essentially rational agents.

We are tempted by the view that every natural causal process admits the possibility of additive interference. The three types of case above are presented as cautionary, illustrating the sorts of situations in which there is an appearance of no interferer for causation. But we say that none of the cases are decisive against our view.

8.4 Isolation and Expansion

We have argued against the inevitability of an effect, given the occurrence of its typical cause; that is, the circumstances under which the effect in question would typically be produced. Context is thus crucial because a causal factor that is present in one set of circumstances might lead to the effect while the same causal factor in a different set of circumstances might not, if it's accompanied by an additive interferer, for example.

This contradicts a common assumption about the nature of causation, which is that the cause and effect must occur together. Whenever we have the cause, we should have the effect, and we would suggest that this is the orthodox view. In subtractive interference, we remove the cause, and then also the effect, hence subtractive interference is no challenge to the orthodoxy. Unlike the subtractive case, however, additive interference counts against the orthodoxy and also, therefore, against the regularity view of causation. A can be typically a cause of B but without B invariably following A. This is because one can have A, plus a further factor I, and then not get B. We thus claim that it is possible for A, considered as a type of causal factor, to be a typical cause of B, another type of factor, but where there are some instances of A, $a_1 - a_n$, that are not followed by an instance of B.

There is, however, resistance to this view. That causation is sensitive to contextual change is considered to be a theoretical problem by both scientists and philosophers. This could be because it is regarded as undermining the claim that A causes B, according to the suggested orthodoxy. For if A does not guarantee B, in what sense does it really cause it?

Scientists typically opt for a strategy of isolating the cause in the lab or in a model, as we saw in Chapter 6, in order to explain how causes are genuinely productive of their effects even though they are susceptible to additive interference. This can be interpreted as a strategy of making the cause smaller and smaller in order to secure the regularity, by excluding everything that might interfere. We could call this strategy *causal isolation*, effectively contracting the cause down to a limited number of factors with no interferer present. From these, is it assumed (rather than argued) that the effect will always follow (see Cartwright 1999: chs 3, 4, Harré and Madden 1975, Bhaskar 1975: 214).

Another strategy, often preferred by philosophers, is to add more context to the cause, such as in Mill's notion of a total cause (Mill 1843: 217). Here the idea is that if you expand the cause so as to include every factor that's in play, then you will get the regularity whenever you have exactly this large set of factors. We call this strategy for saving perfect regularities *causal expansion*.

Again, we suggest, the ensuing regularity in these circumstances is assumed more than argued. The default assumption seems to be always that if there is the same cause, then there is the same effect, and the subsequent discussion centres around acquiring an adequate sense of *same cause* through its precise specification, whether this is big or small (cf. Mumford and Anjum 2011: ch. 3).

The problem with this bigger, expanded cause is that the larger the set of factors that counts as the cause of E, then the less suitable it will be to deliver a regularity of cause and effect, since, as Russell (1913: 8–9) noted, the chance of recurrence becomes infinitesimal. We are likely, instead, to just have a single instance of a large set of factors and where it is followed by E. But then we are left with Hume's old problem of justifying the view that this large, singular event, caused E. All we could have is *post hoc, ergo propter hoc*, the fallacy of assuming E was caused by C just because it followed it.

There is a further problem with causal expansion. If you expand the cause to the nth degree, it becomes the whole world, or at least the whole backwards light cone of the effect. The backwards light cone is the space of everything in the past that could have causal influence on the effect in question. This means that two different effects can be judged as having identical causes, which is uninformative. For example, assume a sphere is caused to both rotate and heat up. These two events, because they have the same spatiotemporal location, have identical backwards light cones. And if that's what counts as their total cause, then they are judged to have the same cause, which means the notion of cause invoked has little explanatory role to play. Having the same cause, there is nothing that informs us why in the first instance a rotation is caused and in the second instance a heating is caused. This shows that if the cause becomes very big, it is no longer of explanatory use or interest.

The idea that causes issue in perfect regularities is challenged by the obvious possibility of interference and prevention. Nevertheless, the thought that regularities should follow from causes, if only they were allowed to do their work, remains persistent.

There are two strategies designed to preserve the idea of causal regularity in the face of the possibility of prevention. We already discussed idealization in Chapter 6, and in this chapter we have introduced the strategies of causal isolation and expansion. But while these approaches might explain how there could be perfect regularities *in theory*, even though there is prevention in reality, none of them prove this. Rather, the strategies involve only the assumption of regularity, accompanied with an alleged explanation of how causal regularity is nevertheless consistent with prevention.

Because these accounts fail to deliver the required regularity, however, it is questionable whether they have succeeded. In idealization, the regularity is only in artificial circumstances and is thus capable of interference as soon as the cause is taken out of its isolation. And with causal expansion, the cause and its effect can become wholly singular and non-repeatable, which is of course unable to deliver regularity. We conclude that the idea of perfect regularity or repetition is not a sound basis on which to rest our scientific methods. Instead, we need a more nuanced account of the worldly order that causal connections deliver so that our scientific methods can look for the proper signs of causation.

9

From Regularities to Tendencies

9.1 Perfect and Imperfect Regularities

We have just considered the issue of interference and seen how a typical cause can occur without its typical effect. In other words, a causal factor that in some set of circumstances or context could be productive of a specific effect, in another context is not. We now need to develop the implications of this point. It may be misleading in its simplicity for it implies, we suggest, major revision to some standard ways of conceiving causation. The revision, however, is one that we think is perfectly acceptable within the sciences and indeed is reflected in scientific methods.

It will be recalled that the constant conjunction theory suggested that, for a real case of causation, whenever the cause A occurs, the effect B occurs. If A occurs without B, in this account, then A is not the cause of B. For causation, we need that A is followed by B in its every instance. We will call this kind of case a perfect regularity. Among the many known difficulties accompanying the view we can add that such perfect regularities are rarely found in reality, even in cases where causation is asserted with some certainty.

For instance, let us accept that alcohol consumption is a major cause of violence globally. There is compelling evidence for this claim (Bellis 2010). But not everyone who drinks alcohol will become violent. And in the cases where alcohol did contribute towards violence, not everyone who drinks alcohol, even in the same circumstances, becomes violent. It is only a proportion of the instances of alcohol drinking that leads to violence. Nevertheless, the causal connection is regarded as sound and well founded. How could one reconcile this empirical fact with a regularity view of causation? The example is not special or rare, we should add.

Suppose we find that 1 in 50 people experience a certain outcome to an intervention. Assume also that this proportion appears reasonably stable over a number of separate studies and that the 1 in 50 incidence is higher than the incidence without the intervention. One might start to wonder whether the intervention caused the outcome. But there is no perfect regularity here. The majority report no effect from the intervention at all. But if the proportion of 1 in 50 reporting the outcome remains reasonably consistent, we might start to think of it as a kind of regularity and in need of an explanation. This is what we would like to call an *imperfect regularity*. It is not that A is always followed by B; but A is sometimes followed by B, in a fairly stable proportion of instances, and which is a higher proportion of instances of B than when A is absent.

An implication of the views we saw in Part II is that imperfect regularities could ultimately dissolve into perfect regularities if only we specified all the facts that were relevant to the effect. So although not everyone becomes violent from drinking alcohol, perhaps everyone in a certain sub-class does, in very particular circumstances. This idea amounts to the suggestion that every imperfect regularity, if it does indicate a genuine case of causation, is an artefact of our descriptive practices. The imperfection of the regularity resides in how we are describing the case, not in the worldly facts of causation themselves. We also saw how a range of similar strategies have been introduced, both in science and in philosophy, to get perfect regularities from apparently imperfect regularities: through causal isolation or abstraction away from interferers, causal expansion to a 'total cause', an assumption of ideal conditions, the addition of *ceteris paribus* clauses to our causal predictions, and so on.

The expectation that there is such a perfect regularity at the bottom of causation is thus a frequent one, even if it is largely tacit in the way we talk about causes. Hence, it could be possible, through adoption of strategies of this sort, to uphold Hume's thesis that causation involves constant conjunction ultimately. If we look to real empirical cases where causation is believed to be present, however, it must be admitted that we do not find de facto pure regularities. Hume's thesis is clearly not backed up with simple data or other forms of evidence. We also considered problems with these strategies. Do they demonstrate perfect regularities or merely assume them without any empirical proof? We will not pursue that question further here. What we will do instead, in this and in Chapter 10 (see also Anjum and Mumford 2018a), is show that there is an alternative way in which causation could be understood.

9.2 Introducing Tendencies

The search for perfect regularities, we argue, could be misconceived from the start. It might be wondered what we could look for, in searching for causes, other than perfect regularities, but we offer an alternative. Instead of causation requiring the existence of Humean constant conjunctions, it is best understood, and sought, in terms of *tendencies*. The idea we suggest, in place of Hume's, is that a cause *tends* or disposes toward its effect, and on occasion succeeds in producing it. In what follows, we will use the term 'tendency' to indicate a variety of related matters. We can speak of a causal tendency or any tendency to cause, such as the tendency of arsenic to cause death. But the term can also be used to name the effects of a cause that tends to happen, for instance the tendency of objects to fall to the ground, which is the outcome of gravitational attraction. A tendency thus also names the type of less than perfect regularity that we think is produced by a cause over a number of instances. Tendency can also refer to a type of modal connection, which we will be discussing in Chapter 10.

While perfect regularity implies that the effect should happen every time one has the cause, at least under some very specific conditions, tendencies don't behave that way. Rather, an effect tends to happen—it is disposed to do so—and will manifest in a less than perfect regularity of the effect, given the cause. Tendencies, we also allow, come in degrees: they can be stronger or weaker. A strong tendency will tend to manifest itself more than a weaker one. We might then try to measure the frequency or proportion of those effects, if they have multiple instances, as a way of ascertaining how strong is the tendency. But a tendency to manifest is no guarantee of doing so, thus there is no necessity that a strong tendency actually manifests more often than a weaker one. There is only a tendency to do so. We should also avoid confusing a strong tendency toward an effect with a tendency toward a strong effect: a small quantity of gelignite and a nuclear reactor can both tend toward an explosion but the latter explosion would be much stronger than the former. Having a strong effect or a weak effect is not the same as the strength of tendency toward that effect; hence the gelignite could strongly tend toward its weak explosion while the nuclear reactor might only weakly tend toward its strong explosion.

What, then, do we mean by a strong tendency toward an effect? Think of our tendency to breathe, for example. Normally this tendency is so strong that we have to actively interfere in order to prevent it, if we ever wished to do so. Breathing is such a strong tendency, it is virtually irresistible. Another such tendency is the gravitational attraction between the Earth and objects on its surface. You may jump in the air but you will soon come back down. We can sometimes produce forces in the opposite direction strong enough to resist gravity, as when we send something up in a hot air balloon. But in almost all cases, the gravitational attraction is so great that objects will naturally rest on the surface of the Earth, or on top of other objects that are on the Earth.

Weak tendencies are less obvious and can easily go undetected, especially as some can be extremely weak. We might be able to detect them statistically, though, in a slightly raised incidence of some effect. Suppose there is a weak tendency of some substance to cause death, for example. Even if it does so only in 1 in 4000 instances, it is likely to be classed as dangerous. It might even become banned. Of course, it is quite conceivable that in this case, the alleged causal link would be contested, precisely because it is not so obvious and thus the statistical evidence in favour of there being such a tendency is open to question.

It can be useful to model tendencies as vectors. The simplest case of a vector model is a diagram with two possible outcomes, F and G, represented on a one-dimensional horizontal 'quality space' (see Figure 9.1). If the quality concerned is temperature, for example, then we could think of F as cold and G as hot. The vertical line is the current situation: the temperature is mild, for instance. This represents the starting point for the causal factors we are considering. Operating upon the temperature are two causal factors, perhaps individuated by the fact that they are causal powers of different objects, and both disposing toward heating, though one more so than the other. We

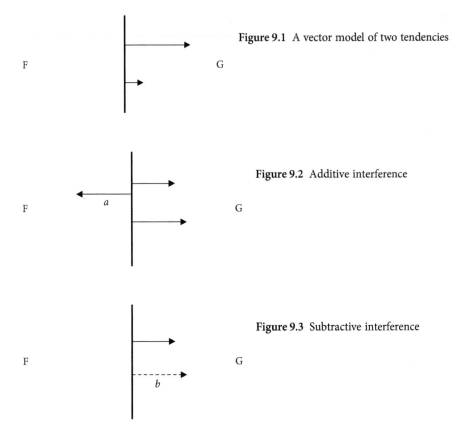

Figure 9.1 A vector model of two tendencies

Figure 9.2 Additive interference

Figure 9.3 Subtractive interference

represent these causal powers on the quality space as vectors. This is useful because, like causes, vectors have a direction and they have a magnitude or intensity, indicated by the vector's length. Similarly, a tendency is directed toward some effect with a certain strength.

The model fits for other quality spaces too. A person can be more or less irritable, one poison could be stronger than another, one substance might emit radiation more so than another, one force might tend toward a greater acceleration than another, one man's sperm might be more fertile than another's, and so on.

Tendencies, because they are causal and disposing toward an effect, are sensitive to contextual interferers. As we saw in Chapter 8, this can be by subtraction or addition. The tendency to breathe can be interfered with subtractively, such as by holding one's breath while swimming under water. It can also be interfered with additively, by a piece of food blocking the throat, or through strangulation. Using vector models, we can also see quite clearly the difference between additive and subtractive interference. In Figure 9.2, a new causal power, *a*, is added to an existing configuration of powers. In Figure 9.3, subtractive inference, causal power *b* is removed.

9.3 Science Deals with Tendencies

We are suggesting that causal regularities in science are better understood as tendencies. Recall that scientific truths are sometimes thought to invoke implicit or explicit *ceteris paribus* clauses, or some very specific conditions under which the cause will produce its effect. This suggests that scientists already acknowledge tendencies, at least when moving from the abstract theoretical sphere and into the real life complexities of open systems. For what better way to understand the *ceteris paribus* clause than as qualifying the causal claim to which it is appended as *having a tendential nature* only?

Consider economics, which is one of the sciences that uses explicit *ceteris paribus* clauses for its laws. One example is the law of demand, stating an inverse relationship between price and quantity of demand. This means that when the price of a product goes up, the demand for that product decreases; and when the price goes down, demand increases. But even though this is called a law, it is never asserted that it amounts to a perfect regularity. If the product is one that some regard as indispensable, and it has no serious market competitors, then demand tends to be stable irrespective of the changes in price. The Norwegian wine monopoly provides such a case. It is owned by the government, which is the sole provider of take-away drinks with a higher alcohol content than 4.75 per cent. Consequently, demand can remain unaffected by a price increase. There are also perverse cases where an increase in price leads to increase of demand, because it makes the product seem more attractive to buyers. People are more likely to buy a mid-priced wine than a very cheap one, even if the two wines are otherwise the same (Janssen and Zander 2014). For many goods and services, price is taken as a proxy for quality even though it is not always so. There are these exceptions, therefore, although they do not invalidate the law of demand if it is understood as a tendency. Indeed, increases in price do tend to reduce demand. It is just that they do not reduce demand in every single case. There is a less than perfect regularity, yet it is a regularity of sorts: a tendential regularity. It concerns what tends to be the case rather than what is always the case. And despite its exception cases, it is a pretty good rule of thumb, capable of providing useful predictions: predictions that tend to be right.

This is an example from a social science and it may be thought that economics is not a typical science. Shouldn't we be looking at the 'hard' sciences? What about physics, where there seems to be more perfect regularities than tendencies?

But we could say in response, first, that it is physics that is atypical in this respect. Sciences such as biology, chemistry, psychology, and meteorology show us how there can be exceptions and interferers, and thus we will find tendencies rather than constant conjunctions. Hence, most sciences will be more like economics than physics. Physics would be the odd one out. Second, the view that physics really does reveal perfect regularities is in any case contentious. As we have seen, the perfect regularity is to be found in idealized circumstances or abstract models, which are thus artificial. As soon as we take physics into the 'real world', we find that physical phenomena can interfere with each other just as much as those in any other science (Cartwright 1999). Consider the movements of bodies according to dynamics, for example. They frequently get in each other's ways, have their direction and momentum changed through collision, and so on. Nevertheless, we claim, one can still detect tendencies in physics, such as attractions and repulsions; and thus the tendency account fits here too.

9.4 Is a Tendency Nothing but a Statistical Incidence?

We have argued that tendencies differ from perfect regularities, where there is a constant conjunction of cause and effect. Instead, the only type of regularity causation produces is a tendential one. What does this mean for scientific methods?

Perhaps tendencies could be associated with, or even inferred from, statistical incidences? A way to detect a tendency would then be to look at the frequency or proportion of the outcome in a given sample of instances where the cause is present. The tendency of X to cause Y could then be the same as the proportion of Y within an X-sample. If Y occurs four out of ten times when there is X, then the incidence of X among Y is 40 per cent. Any such statistical method requires, of course, multiple repetitions of the cause, which may not always be available to us, perhaps because the cause is a rare occurrence and one beyond our ability to manipulate.

If this is all we mean by a tendency, then statistical results and quantitative studies alone would help us reveal them. But we should be cautious about drawing this conclusion. Tendencies are not the same as statistical incidences, nor strictly derivable from them, for at least the following reasons:

i. Correlation is not causation. Statistical correlation alone cannot determine whether or not there is causation, and thus cannot determine whether or not there is a tendency to cause either. As shown in Chapter 4, a statistical correlation of A and B could mean at least five different things: A causes B, B causes A, A and B have a common cause, both A causes B and B causes A, or there is no causation at all. When we say that tendencies come in degrees, from very weak to very strong, we are therefore not suggesting that any old raised incidence, in the statistical sense, would amount to the presence of a causal tendency. A raised incidence that is detected statistically may be due to other reasons than simply A being the cause of B.

ii. Tendencies can vary individually. While a statistical incidence might reveal an average tendency that is in fact causal, it cannot reveal individual tendencies. For instance, even though on a population level the frequency of an effect, given the cause, is 40 per cent, this does not mean that in each individual case, the tendency is exactly the same. If we are talking of an incidence, then we are looking at an effect averaged over a population. One cannot, therefore, infer straight from incidence to the presence of a tendency in a member of the population or set of cases.

iii. A tendency can be unique. There are some cases of causation that only happen once. This does not mean, however, that there is no tendency toward the effect. As already discussed in Chapter 5, repeatability is a requirement for Humean causation, since no regularity can be found without it. The same problem applies to statistical incidence. Suppose an asteroid impact 66 million years ago caused the Cretaceous-Paleogene extinction, wiping out a wide range of species including the dinosaurs. This should still count as a causal tendency of the impact even if the event was not, and will not be, repeated, for all we know. That may seem like a special case. However, in medicine, individual variation—which can amount to uniqueness—should be expected, since each individual will have some causal factors that are unique to them, such as genetics, medical history, diet, lifestyle, relationships, and so on. Uniqueness should not tempt us to deny causation, but instead it shows the important role of causal theories.

iv. Low incidence I: significant but rare. Some tendencies are strong in the individual case, yet too rare to be detected statistically. Since a population sample can only be of a certain finite size, extremely rare effects are unlikely to be uncovered in a study. Statistically, this would tend to show up in a very low incidence, for example, if an effect occurred only in 1 in 100,000 cases. But we should not infer from this that the causal tendency itself is weak. An example of this is a rare but well-understood response to vaccines. While it is common to get a reaction to a vaccine, caused by the antigen injected, an entirely different type of response is anaphylaxis, which happens when there is an allergic reaction to one or more of the components in the vaccine. This is extremely rare and happens in less than 1 in a million cases (Zhou et al. 2003, Erlewyn-Lajeunesse et al. 2007). Because anaphylaxis from vaccination is such an uncommon event, not a single case was detected even in the largest clinical trial, which also makes it impossible to estimate the incidence of this effect scientifically. But even if it could be estimated, we might still conclude that the tendency itself is so weak that it is negligible. It would surely be a mistake to neglect the possibility, however, since anaphylaxis is a 'severe, life-threatening, generalised or systemic hypersensitivity reaction' (Erlewyn-Lajeunesse et al. 2007). In the individuals with the allergy, the tendency is so strong that precautions are advised to be taken every time a vaccination takes place just in case the effect occurs.

v. Low incidence II: insignificant but frequent. There are some tendencies that are so weak that they never produce their effects, but the tendency is still there in each case. One example is the cyanide found in apples. Apple seeds contain amygdalin, which is a poisonous cyanogenic glycoside (sugar molecule). While cyanide can be lethal at as low as 1.5 mg per kilo of bodyweight, the amount of cyanide in apples is too small to ever kill a person, ranging from 1 to 3.9 mg g^{-1} (Bolarinwa et al. 2015). Statistically, this case would look the same as under *iv*, but there is an ontological difference between the very weak tendency of the apple to poison and the rare but very strong tendency of the vaccine to anaphylaxis. While the cyanide is intrinsically a part of each apple seed, the allergy is only intrinsic to some very few individuals.

Searching for tendencies purely statistically must therefore be done with a degree of epistemic humility. The idea of epistemic humility is that there are some truths that it is not possible to know with any high degree of certainty no matter how much data one has. A statistically raised incidence of some phenomenon might not be an accurate reflection of the real nature of the tendency. In other cases, a tendency might not be statistically detected or detectable. Detecting and sometimes counteracting weak tendencies is important, if outcomes are severe, but despite our best efforts we might still not know of them, especially if we limit our enquiries to statistical data. Consider a relatively weak tendency of a particular potential cause but which thus far has happened only once. Further, assume that on that one occasion, the weak tendency failed to produce its possible effect. It might then be assumed that there is no such tendency at all. Yet this is consistent with it indeed producing the effect if a second instance of the same tendency were ever to occur.

9.5 Tending away from Perfect Regularities

We advocate moving away from thinking about causation in terms of perfect regularities and start looking for tendential, less than perfect regularities instead. A tendency is still a type of regularity, but not anything like Hume's constant conjunction, where a cause is always followed by the effect. A tendency is what tends to happen.

At times we have ventured into matters of underlying ontology too. We have spoken of a tendency as being the thing that produces the less than perfect regularity. Perhaps this is a conflation of that which produces the regularity with the regularity that is produced. Indeed, one can see how both these elements could justifiably be called the tendency. But we also hope the distinctness of the elements can be discerned. There is first the causal factor, which may or may not produce its effect in any of its instances. There is then the effect which that causal factor tends to produce. And, as we have explained, there are also the statistical facts of incidence, available to us when there have been multiple occasions on which the cause occurred, with some proportion of cases where the effect was produced. It is these facts of incidence that we have called the tendential regularity: a pattern of events, though one likely to be less than constant. We also argued that any such regularity, while being a product of an underlying tendency, could not be used straightforwardly to infer the tendency and, in particular, that the incidence of occurrence of an effect did not automatically indicate the strength of the tendency involved.

10

The Modality of Causation

10.1 Thesis: Hume against Necessity

As well as something being the case, it could also be necessary. That $2 + 2 = 4$, that the past occurs before the future, and that men are mortal, are all examples of necessary truths. On the other hand, something could be not the case but nevertheless possible. It is currently false that the Statue of Liberty is painted red but it seems at least possible that it be so. If the people in authority had the will to paint it, they could make it happen. Whether something is necessary or possible is known as its modality or modal value (see Melia 2003 for a survey). Questions of the modality of causation naturally arise. When the cause occurs, for instance, is the effect then necessary? Does a cause produce its effect through necessitating it or does it merely make it possible? We will argue here that both these alternatives are mistaken, and that the modality of causation is instead something in-between them.

Hume has again been influential in this discussion. He presented causation as a relation between objects or events A and B that involved three characteristics (Hume 1739, I):

1. *Constant conjunction* of A and B (whenever A, B).
2. *Temporal priority* of A over B (A before B).
3. *Contiguity* of A and B (A spatially next to B).

What his analysis excluded, controversially, was some underlying necessary connection between A and B. Hume denied that when A was the cause of B, A also necessitated B. We will examine the justification for his view, considering its merit, and what implication the view has for questions of scientific method.

In considering the reasons Hume held this anti-necessitarian view, we must understand his commitment to empiricism. Empiricism is sometimes expressed as the view that all knowledge concerning matters of fact originates in experience. Hume (1748: 62) put this more elegantly: 'all our ideas are nothing but copies of our impressions, or, in other words, that it is impossible for us to *think* of any thing, which we have not antecedently felt, either by our external or internal senses'. Hume conceded, however, that we sometimes mistakenly believe we have an idea that has originated in experience when really it hasn't. The necessity of causation is his main example. We often believe, or say we believe, that the cause necessitates the effect. But

Hume challenged anyone to show the original experience from which we could know of any necessary connection. He believed that there is none. We cannot therefore know that there is a necessary connection between causes. Indeed, we cannot even know what we mean by such a connection.

Hume gave detailed arguments for this view. He said that we cannot know of any such necessity in nature a priori; that is, using our reasoning alone. Nor can we know of necessity from what we experience. The following gives us a flavour of Hume's thought:

Let an object be presented to a man of ever so strong natural reason and abilities; if that object be entirely new to him, he will not be able, by the most accurate examination of its sensible qualities, to discover any of its causes and effects. Adam, though his rational faculties be supposed, at the very first, entirely perfect, could not have inferred from the fluidity and transparency of water that it would suffocate him, or from the light and warmth of fire that it would consume him. No object ever discovers, by the qualities which appear to the senses, either the causes which produced it, or the effects which will arise from it.   (Hume 1748: 27)

We believe that water causes drowning, therefore, only through experience. But what is the nature of that experience? Hume goes on to analyse it, in general terms, as nothing more than constant conjunction coupled with temporal priority and contiguity. We have experience of other people and ourselves being in water and then unable to breathe. This is nothing more than experience of a regular pattern of events. Crucially, no matter how much one looks, one will never find an experience of any necessity between cause and effect.

In the first place, one cannot find necessity if one examines a single instance (Hume 1748: 63). All one finds there is one event followed by another. If it were not for seeing that the first type of event is followed regularly by the second, then one would never get the idea that the first caused the second, let alone that it did so through necessitating it. If we have repetition or constant conjunction, Hume says, then we can get the idea that A caused B (as long as the other conditions are met of temporal priority and contiguity). But Hume argued further that seeing the repetition of A followed by B was never enough to generate the idea that A necessitated B: 'From the mere repetition of any past impression, even to infinity, there never will arise any new original idea, such as that of a necessary connexion; and the number of impressions has in this case no more effect than if we confin'd ourselves to one only' (Hume 1739: I, iii, 6, 88). This argument has wide-ranging implications. Hume thought that when we say that A causes B, we can mean nothing more than that A is constantly conjoined with B, temporally prior to and spatially contiguous with it. But, if that is the case, we cannot say, for example, that B happens *because of* A; only that whenever A happens, then B happens. Because any notion such as real causal force, power, energy, and necessity cannot be known, according to Hume, the world he describes is one of *pure contingency*. By this we mean that, in theory, anything can follow anything. There is this kind of possibility, but there is no necessity in nature.

It may seem relatively easy, and indeed natural, for science to adopt this way of thinking. It looks conducive to the idea of science to say that all the components of a causal judgement are perfectly empirically accessible. There was thus the movement known as logical empiricism (some say logical positivism, see Hanfling 1981), which could be interpreted as the application of Humean philosophy to scientific method. It is possible to know that A is always accompanied by B, always before it in time and next to it in space. Causation is thus reduced to three empirically knowable relations. Given knowledge of these, then science would know every causal connection, just by gathering the data on the three relations. The methods of science, when seeking causal connections, are just to look for these facts. In contrast, were one to claim that there was a necessary connection involved in causation, one would be effectively engaging in metaphysics—going beyond experience—which Hume argued was ultimately meaningless and should be consigned to the flames (Hume 1748: XII, iii, 165).

There are, however, difficulties if one aims to build a science on this kind of strict empiricism. Among other issues, a world of pure contingency leaves us with the problem of inductive scepticism. Given what has happened in the past, how can we know that a future A will be followed by B? Experience must concern only the past and present known and recorded As. It says nothing about the unknown and future As. And, without that, how do we have any good reason for action where it concerns a new case? Hume admitted that every known A being followed by a B tells us nothing whatsoever about a future A: not even that it will *probably* be followed by a B. What, then, would be left of prediction or explanation? Science would seem completely unable to tell us about future and unencountered cases and thus would be able to tell us nothing about what intervention is warranted. It cannot even judge that it is rational to eat food. The fact that food has nourished us and other people in the past says nothing at all about whether it would do so in the future, at least according to Hume's argument. And thus it is not strictly rational to eat food. Hume saw that this was a major problem for his theory (Hume 1739: Conclusion of Book I). He did not glory in his findings but saw them as deeply problematic. Nevertheless, he could see no rationally defensible alternative, nor any flaws in his reasoning.

10.2 Antithesis: Necessitarianism

Hume effectively denied that there was any real modal force to causation, certainly as anything more than the pure possibility with which anything can follow anything. There was nothing in a cause that required, compelled, or made the effect happen. If the world is as Humeans think it is, then everything merely is the case or is not the case: it does not have a modal value besides. To state the issue in this way would be to equate modality with necessity, which is often done, but we will argue that we need more refined modal distinctions. For one thing, being false but possible is an ascription of a modal value: of possibility. But traditional Humeans don't accept the reality of unrealized possibilities so are at liberty to say that everything just is or isn't. We cannot ascribe any qualities to what isn't, and calling that which already is the case *possible* is redundant, so possibility effectively disappears from a Humean view.

However, because of the problem of induction, and finding the analysis of causation to be implausible, many seek an anti-Humean account of nature. They reject the idea that ours is a world of pure contingency or mere possibility in which anything goes. Some anti-Humeans argue, on the contrary, that necessity is essential for causation and that this solves the problem of induction (e.g. Bhaskar 1975: 206–18, Ellis 2001: 283–6). Their idea is that a cause genuinely produces its effect and it does so through necessitating it, in some way (Ellis 2001: 7).

Suppose Hume is right and there is no direct experience of necessity in causation. Suppose also that we are sympathetic toward the underlying idea of empiricism: that knowledge concerning matters of fact must come from experience. Even then one needn't follow Hume in applying this standard so strictly. Perhaps there are some broader considerations concerning our experience of the world that we ought to take seriously. For instance, although you can never see that a cause necessitates its effect, you can see that science is broadly successful. It makes predictions that, while not 100 per cent accurate, are accurate and reliable enough that it has enabled us to make new materials and drugs, build and fly aeroplanes, predict the exact times and places of solar eclipses, develop computer technology, and so on. When a precise prediction is subsequently confirmed, it is quite a validation of science and the idea of empirical knowledge generally. Can we really say that this success could happen if the world was as Hume told us it was, where anything could follow anything? There is a no-miracles argument that we can invoke here (Putnam 1975: 73). The success of science cannot be tantamount to a miracle. And how can we explain the success of science unless causal knowledge was useful, with a track record of satisfied expectations? Surely this tells us that causation is real and that it involves a stronger modal connection than mere possibility.

If a cause necessitates its effect, then we have a basis on which to believe that future and undiscovered cases will be like past recorded ones. Defenders of necessity will reason like so. If A has always been followed by B, in some cases—perhaps not all—there's a reason for this: namely that A has causally necessitated B. But if A necessitates B, then we have a reason to believe that any future cases of A will also be followed by B. This would be an instance of a more general counterfactual truth that if A had occurred, then B would follow. Hume's account could not sustain such counterfactual reasoning.

Despite the need for an alternative to Hume's world of pure possibility, and any attraction to necessity as a way of solving the problem of induction, there are nevertheless problems with this alternative too. One might, then, think of it as an overcompensation for the problems in Hume's account.

In the first place, where would such worldly necessity come from? One alternative is to ground causal necessities in the governance of laws of nature (see Chapter 18). Nomological or lawlike necessity would be the idea that there were strictly necessary laws, of which particular causal transactions were the instances. For example, there is a theory of laws due to Dretske (1977), Tooley (1977), and Armstrong (1978) in which a law relates two properties, F and G, when they stand in a relation of natural necessitation, which Armstrong represents as N(F,G). There is a natural necessitation between the properties of being a heated iron bar and expanding, for example. The idea is that nomological necessity entails a constant conjunction but is not entailed by a constant conjunction: there could, after all, be accidental regularities that were not causal. The difference between a regularity that is accidental and one that is genuinely causal is that the latter instantiates such a law of nature while the former doesn't. This could give us everything Hume wanted from causation—constant conjunction, temporal priority, and contiguity—but, in addition, it gives us necessity too. There has been much discussion of whether this is a plausible account of how the natural world contains necessity (see Mumford 2004: ch. 3).

However, setting that discussion aside, there is perhaps an even bigger problem with this, and any other necessitarian account. Isn't it too strong? We have seen, over Chapters 8 and 9, grounds to question the assumption that there is ever a constant conjunction in nature, given that causal processes seem subject to additive interference (Chapter 8) and the regularities we find in nature seem less than perfect (Chapter 9). Anti-Humeans appear to accept Hume's claim that causation involves constant conjunction but add that it must involve *more* besides, namely necessity. We could think of this as a 'constant conjunction-*plus*' strategy, with the idea being to find the extra ingredient that accompanies constant conjunction when we have genuine causation. But there seem to be grounds for thinking of causation as involving something *less* than constant conjunction, according to the arguments we have presented.

Are Hume's opponents over-reacting to his thesis that causation is purely contingent by claiming, implausibly, the antithesis: that the causal is necessary? But if we think so, it seems we are saying that both Humeanism and anti-Humeanism are wrong. How is that so? Is there any alternative or do those two positions exhaust all the possibilities?

10.3 Synthesis: Tendencies

Between the pure contingency of Humeanism and the necessitarianism of some anti-Humeans, there is, thankfully, a third view. While it has its origins in the ancient and medieval philosophy of Aristotle and Aquinas, it is a view that has had a recent revival (Mumford and Anjum 2011: ch. 8). In Chapter 9, we explained how the world seemed to have some kind of regularity but it was less than perfect. We said that this was a tendential regularity, where some things tend to happen.

We can now provide a theoretical background to this idea in relation to the matter of modality. Specifically, we wish to defend the view that the modality involved in causation is neither pure contingency nor necessity but something in-between. This is sometimes called the *dispositional modality*, considered to be a modality stronger than mere possibility but weaker than necessity. Causes tend or dispose toward their effects, such that they sometimes successfully produce them but need not always do so, and this is why there are tendential regularities in nature, exhibited in raised incidences, for instance. Furthermore, such tendencies can come in different degrees of strength, from very weak to very strong. A strong tendency toward some effect might dispose toward a high incidence of it, and a weak tendency might dispose toward a relatively low incidence of an effect, though perhaps still higher than if there were no tendency at all (see Section 9.4 for a discussion of weak tendencies).

Note, however, that there is still 'only' a disposition toward such an incidence. There is no guarantee that a certain proportion or distribution of outcomes is produced. In terms of scientific method, this shows us why the strength of a tendency cannot be read straight off from a frequency of occurrence. For example, if 60 per cent of As are followed by Bs, one cannot infer that As have a 60 per cent disposition toward B. This is for a whole host of reasons, but one is that any such disposition will only tend to produce its effect in a certain proportion of instances. Over a sequence of trails, that proportion might be unrepresentative of the true strength of the disposition.

A possible disadvantage of this account is that it would pose a radically new challenge to some standard ways of thinking. It is orthodox in philosophy to think that there are only two modal values: necessity and possibility. Here would be a third, which is something in-between the two others. So, for instance, a sperm that meets an egg will tend to fertilize it, but there is no guarantee that it will do so. Still, it is more than a mere possibility that it fertilizes it, since there is a disposition for it to do so. And a disposition, no matter how strong, can be counteracted by others, so it always falls short of necessity.

Opponents might nevertheless question whether we need to posit a whole new, third modal value to explain such facts. Furthermore, we find that many who use the language of tendencies, on closer inspection seem to hold a different, *conditional necessity* view (Bhaskar 1975: 14, 50, and 98, Aquinas 'Principles of Nature': §26, Geach 1961: 103, 104, Aristotle *Physics*: II, 8, 199b15–19, *Metaphysics*: IX.5, 1048a15–20 and IX.7, 1049a5–11). Causes are capable of being prevented, they allow, but when all the conditions are exactly right, *then* the cause necessitates the effect. This makes the tendential nature of causes a purely *external* matter to the cause. Less than perfect regularities are produced only because of the dispositions of things interfering with each other. But, one could well ask of a conditional necessity view, does it make sense to say that A necessitates B but only when it is in the right circumstances? It sounds like a reasonable test of what it is to be necessary that something is the case no matter what the circumstances; and thus this is a test that causation fails.

It is possible, however, to hold a *deeply tendential view* of causation instead, which is what we suggest here (for details, see Anjum and Mumford 2018a: ch. 1). What we mean by this is that even when all the conditions are 'right', the cause still only tends toward its effect, which can thus fail to happen. This makes the tendency of a cause an *internal* modal feature of it. In this case, it seems like we have an irreducibly dispositional view of reality.

Thus far we have used the terms disposition and tendency as if they were interchangeable, which might cause confusion. In ordinary language, they may well be equivalent but nevertheless it would be helpful to be clear on the position we are describing. In what we think is the best version of the tendency view, it will be based on an ontology that accepts the reality of causal powers in things. A causal power is a feature of a particular that disposes toward an effect or manifestation. Causal powers are sometimes also called dispositions and they will tend toward their effects in the way we have described, with dispositional modality. Typical dispositions are fragility (the power to break), elasticity (the power to stretch), flammability (the power to burn), charge (the power to attract or repel), and solubility (the power to dissolve). These powers will tend to manifest under certain circumstances but, as we have seen, it is not necessary that they manifest.

We can claim that there is an indirect argument in favour of the tendency view. Even if anti-Humeans were wrong to say that causation involves necessity, and in that respect they don't have a credible solution to the problem of induction, they were right that induction is otherwise a problem. This is to say that the problem of induction doesn't go away just because their solution to it fails. However, we can now say instead that to make sense of the rough success of inductive inferences, you need tendencies rather than necessities in the world. And it is arguable that the tendency view fits the facts more: namely, that things tend to occur, with some degree of incidence, such that good predictions will tend to be right—will tend to be reliable—but are nevertheless fallible in that they may not be borne out in events. Hence, Mill (1843: III, x, 5), Peirce (1892), Bhaskar (1975), and others think that a tendency account is perfectly fit for science.

If one is looking for the empirical signs or symptoms of causation, therefore, one is best to look for tendencies—in raised incidences, for example—than to look for perfect regularities. The latter, after all, could be more symptomatic of something that is not causation, such as identity, classification, or essence, which admit of no exceptions. Tendencies are, rather, characterized by four features: i) they are weaker than necessity but stronger than pure contingency, thus involving a *sui generis*, irreducible modality; ii) they have a directionality toward a particular outcome; iii) they have a magnitude; and iv) they can be counteracted by other tendencies.

A reader might feel disappointment that we have only characterized the dispositional modality without actually defining it. This is true but we need make no apology for it. As we hope has been made clear, we take the modality of causation to be a primitive matter. In particular, we cannot define it in terms of the other two, traditional modal values (in the way, for instance, that possibility and necessity are interdefinable with negation, so that *possibly* can be defined as *not-necessarily-not*). Nevertheless, we have tried to show that even if no definition can be given, hence no reductive analysis, we have offered an informative theory of dispositional modality in terms of its chief characteristics.

10.4 Does It Matter for Science?

Philosophers have traditionally understood modality in a binary way in which there were only two modal values and thus everything was either necessary or contingent, the latter being understood to mean merely possible. It followed that discussion of the modality of causation has largely been confined to this division, assuming a thesis we call modal dualism. Humeans thought that there was no necessity in natural causal processes and this provoked a reaction in which it was claimed that causes necessitated their effects.

In Part III, we have offered an alternative to modal dualism and shown its implication for the understanding of causation. Causes do not necessitate their effects, we say. The possibility of prevention and interference shows that causes do not guarantee their effects. But we have also argued that there are tendencies in nature that need some explanation. There is more than pure contingency in the workings of the world so, in that sense, Hume is wrong. Finally, we have tried to reconcile the ideas of causes involving less than necessity but more than pure contingency. This is, we have argued, because there is a third, and more important, modality at work. Causes tend or dispose toward their effects and this is what sometimes produces the less than perfect regularities measured as raised incidences of some effect.

This view has important implications for scientific method. A prevalent assumption that we examined in Part II is that causation should produce perfect regularity, otherwise known as constant conjunction. Were that so, the job of science would be to find those constant conjunctions. But there will be none: at least none that are symptomatic of the presence of causation. Instead, we need methods designed to identify the tendential regularities that are symptomatic of causation. But this also explains why methods for the discovery of causes are no simple matter. If constant conjunction always accompanied causation, that would make scientific discovery of causes relatively straightforward. Instead, we think that causes only reveal themselves in symptoms such as raised incidences, where one cannot automatically infer a cause from the statistical evidence. At least, then, our account fits the fact that discovery of causation is no easy task.

PART IV

Causal Mechanisms

11

Is the Business of Science to Construct Theories?

11.1 Against Strict Empiricism

There is a short and a long answer to the question above. The short answer is yes: of course, it is the business of science to construct theories. How could anyone think not? While some might find this a perfectly satisfactory response, we will here present the slightly longer one.

It is a legitimate norm of science that we ought to construct theories, we argue, and there are good reasons to say that science simply cannot do without them. This stance has, however, been challenged by scientific trends found in epidemiology and evidence-based policy, but also by philosophical positions such as logical positivism. According to strict empiricism, of which logical positivism was a modern reconstruction, anything that goes beyond the data is speculative metaphysics, and therefore unscientific. On this view, scientists should stick to what they can know through observation and resist the temptation of causal theories. This would make science a purely empirical matter, based on experience alone. The type of experiences with which science is concerned are the data, which can be verified, checked, and supposedly are objective. By restricting science to what is there in the data, it would remain free from metaphysics.

Although there is this positivist tradition within philosophy and science that emphasizes the importance of empirical data and downplays the role of theory (as discussed in Chapters 1 and 2), we can still ask whether this is a defensible approach to science. Could data alone really give us everything we need for scientific purposes? Or would that be to ask too much of data?

Here, in Part IV, we develop further our positive account of scientific methodology. We begin in this chapter by establishing the need for theory in science. But a further question is then raised about how exactly theory relates to data. If we collect an adequate amount of data, do data generate theories? We deal with this question in more detail in Chapter 12, where we discuss whether collecting more data of the same type is the best way to establish a causal theory. In Chapter 13, we move on to argue that scientific theories are better when they involve mechanistic, qualitative, and causal information rather than purely quantitative data. But, as will be made clear in

Chapter 14, our understanding of causal mechanisms carries no commitment to reductionism, since the relevant mechanisms could even be found on a relatively higher level than the phenomena they explain.

By emphasizing the importance of causal theory and mechanistic understanding, this part takes us a step further away from the strict empiricist approach to science, including the Humean notion of causation. Without a theory that can explain why and how a cause is linked to its effect, the data seem of restricted use to science.

11.2 The Alleged Roles of Data

Empiricism is attractive insofar as, without data, science would become pure specu-lation. But should science restrict itself to data? To impose this restriction would leave quite a substantial job for the data to do, since they would have to deliver everything we want from science. Perhaps they can. We will now take a look at some of the things that we might want the data to provide and see whether a data-only science is a sustainable ideal.

Data are thought to be crucial to science because they allegedly *uncover facts* about reality. Some might even say that data simply *are* facts, so that when we collect data we are actually collecting facts. Statistical data, for instance, seem a lot like pure facts. From a quick search in the online database of *Statistics Norway*, we can learn that twenty-two people died in traffic accidents in July 2016, that Nora and William were the most popular names registered in 2016, and that 27 per cent of households include children. These are all very fact-like data.

We can also *uncover relational facts* through statistical data if we introduce comparisons or associations. Co-variance is one type of relation, in which an increase or decrease of one type of factor is correlated with the increase or decrease of another factor. One might find that access to green areas is correlated with good health, for instance. But restricting ourselves to the facts, we should not be tempted to conclude from the data that one is a cause of the other. The types of facts uncovered by data are not restricted to statistics, but could require experimentation or other types of intervention. In a lab using the right kinds of tools and tests, one might discover toxins in the drinking water or in a plant. This still counts as uncovering facts since the toxins were already there before the test, unlike cases where an intervention affects or even produces new facts.

Once we have collected the data, we can use them to *generate hypotheses*. A certain data set might indicate a hypothesis that we would not even think of if it weren't for the data. That smoking is linked to cancer is one example of a hypothesis that was first discovered as a statistical correlation before any of the causal mechanisms were known. Without data, we probably would not even think to investigate this link. Data thus seem useful and perhaps even necessary for finding plausible hypotheses. Suppose we find that married men live longer than unmarried men. One might think there is something about being in a caring and stable relationship that benefits men's health. A further question is then why women are not affected by marriage in the same way as men. Is it because their lifestyle is healthier? Others might turn the explanation around, by hypothesizing that men are chosen as mates based on their genetic advantages. If so, these men do not live longer because of their marriage. They are married because they live longer.

Hypotheses are thus attempts to *explain* the facts. But the data are needed to indicate where we should look for causal connections and theories thereof. It seems, then, that unless we start with data, we would not even reach the stage of attempted causal explanation.

A further motivation for gathering data is if we have arrived at some research question or hypothesis and need to test it. Such *theory testing* is a typical task for data. Suppose a certain teaching method is the norm in academic institutions because it is assumed to have a learning advantage (e.g. extensive use of information technology, including slides, clickers, and smartphones). Then this hypothesis could be tested to see whether it is backed up by data. If it is not supported by data, a revision or rejection of the original hypothesis might be needed. When a hypothesis is supported by data, however, we might say that the data *provide evidence for the hypothesis*. To offer data that support our hypotheses in this way is crucial in most empirical sciences, and it is an explicit requirement of evidence-based policy that we do so. The type of data that counts as evidence for a hypothesis will vary depending on the scientific discipline and its preferred methods. Some hypotheses require experimental evidence, while others might require observations or calculations.

When the data reveal that something happens regularly and over various contexts, we might use them to *make predictions*. If there is substantial data showing an increased learning outcome from use of smartphones in teaching, one could predict that the same outcome would occur for other students and teaching situations elsewhere. This prediction would then be grounded in data, and the more data one has, the more confidence one might have in the prediction. That data can ground predictions in this way is crucial for evidence-based policy, where data are systematically used to *inform policy decisions*, for instance about which teaching methods should be used in all academic institutions.

It seems, then, that data can indeed provide a lot of what we want from science:

- uncover facts
- indicate hypotheses
- explain
- test hypotheses
- provide evidence
- ground predictions
- inform policy decisions.

If this is correct, do we really need to offer theories over and above data? Possibly not. We should be cautious, however, of accepting uncritically the view that data could play all these roles without further qualifications. Despite the positivist ambition of a metaphysics-free science, and the empiricist ideal of experience-based knowledge, the relationship between data and theory is not as clear-cut as it might seem at first glance. If taken entirely on their own and in isolation from a theory, data might not be able to do any of this work we want from them, as we now go on to explain.

11.3 No Explanation or Prediction without Theory

One of the main motivations for doing science at all is exactly our need for explanation and prediction. We want to understand what goes on around us, but also to predict what will, could, or might happen next. Without explanation or prediction, any choice of action would be entirely unmotivated. The success of science thus lies in its application and utility for these purposes. Only to the extent that science provides explanation and prediction will we be able to use the knowledge we get from it to affect or change our decisions about what to do. And any policy recommendation is, implicitly at least, motivated by the expectation that what happens in one or many studies would also happen elsewhere (see Cartwright and Hardie 2012 for a critical discussion on this). How could data alone justify this claim?

At best, the data tell us what happened. But they cannot tell us why it happened, nor what would, could, or will happen in other cases. To get from data to the other business of science—explanation and prediction—we need a theory, and perhaps even a causal theory. We might say that it is the data themselves that are in need of explanation.

Consider the following example. We might observe a high mortality rate and a raised incidence of a bacterium. From that, a theory could be constructed that relates the bacterium and mortality: that the bacterium is sometimes lethal. This is a causal theory which can then be used to explain the raised mortality in terms of the bacterium. It could also be used to predict that if someone is exposed to the bacterium, they might have an increased risk of death. The reason why theories can be used for these purposes is exactly that they go beyond any specific instances. In order to give us any information about how to use the data, we need to have some further idea of how they relate to cases that are not themselves part of the data.

Theories are problematic for the empiricist because they involve general claims, but without this feature, theories fare no better than the data. One might know that person a had the bacterium and died and that person b did too. But the theory is not about any specific person or persons. Theories are general because they are abstracted from and independent of particular individuals and instances. By suggesting a potentially lethal quality in the bacteria in general, the theory allows counterfactual reasoning about what might happen in other instances that are not yet observed: *if person c who is not currently exposed to the bacterium, were to be so exposed, then c would be at risk of death.*

Unaccompanied by a general causal theory, the data themselves would remain impotent. They would have no application beyond the particular instances that make up the data set. Such generalizability of data is called *external validity*, and without it, no data set would ever suffice to draw conclusions about what might, could, or should happen elsewhere.

It seems that the strict empiricist ideal of science is not easy to defend without considerably limiting the role of science. A significant part of science is non-empirical, forcing us to go beyond the data. It might not be easy to detect when exactly we are making a non-empirical claim or assumption. But even in asking whether or not our data have external validity, we admit that there could be scientific reasons to draw conclusions beyond the data. To draw a general conclusion based on data alone is to make an invalid and inductive inference; unless, of course, data actually dictate scientific theories. Can they do this? Or is this another empiricist ideal that needs to be questioned? We say it is.

11.4 Do Data Dictate Scientific Theories?

We saw that science needs theory to explain and predict, but that it can only do so if those theories are general. How does such a theory relate to data? If theories are general and data are about particular instances, then a purely empirical approach to science requires that data can somehow give us the correct theory. This would be a major vindication of data and, indeed, of the theories that come from them. It would also mean that science would be a relatively simple process of discovery.

Historically, this type of claim has been made by Francis Bacon (1620) in his influential work *Novum Organum*. He there introduced a scientific method that is supposed to take us directly from data to theories. First, we start with a set of single observations, without any expectations about what to find. Only when we observe free of dogmas, will our data be raw and uninfluenced by theory. It is also important at this stage that the observations are done methodically and repeatedly and under a variety of contexts. Once the data are collected in the right way, we can use them to form reliable general claims. We now have a hypothesis or a theory that can be tested with the necessary tools. The results of the test are finally used to confirm or reject the theory. On this inductivist approach, the data generate theories 'like a machine'. The crucial step of the method is the data collection. After this, the hypotheses should present themselves to us almost automatically.

Bacon's inductive method faces a number of well-known problems, however. One is that a set of data is consistent with many different scientific theories, which means that there is always more than one theory that could fit and even explain the data equally well. This is called the problem of underdetermination and examples are easily found. Based on the exact same data, but because of different theories, the Aristotelians saw the Sun rising while the Copernicans saw the Earth rotating. And the different approaches to evolution posed by Lamarck and Darwin are not disagreements over data, but over the theories explaining those data. We can also understand underdetermination mathematically, as Russell (1948: 496) did when he said: 'The difficulty is that, given any finite set of observations, there are always an infinite number of formulae verified by all of them.' This point is visualized in Andersen and Becker Arenhart (2016: 169), where a set of data points are plotted on a graph. Usually, we will draw a straight line through them that connects them and fills in the missing values, but there is in principle nothing stopping us from instead drawing a zig-zag pattern between the points. Both these lines would be consistent with the data, so the data cannot determine which is right.

If the data do not dictate the theory, perhaps they could still bear some other intimate relationship to it. Shouldn't any theory be at least consistent with, and thus restricted by, the data? This is the view of constructive empiricism. Van Fraassen (1980: 12) argues that scientific theories must be empirically adequate in the sense that they 'save the phenomena'. Constructive empiricism maintains the positivist ideal that science should be free of metaphysical claims, while also conceding that science is incapable of uncovering truths beyond the observable (Monton and Mohler 2014). Acknowledging the problem of underdetermination, constructive empiricism takes instead an instrumentalist approach to theory: one should choose the theory that has the greatest instrumental value among all those that are consistent with the data.

This might seem like a minimal requirement on scientific theory: that it be consistent with the data. But even this is not straightforward. There could be circumstances in which we are willing to accept a theory even though it has an inconsistency with at least some of the evidence. Lakatos (1970) argues that theories are 'born refuted'. Rather than simply stating accepted facts, he says that scientific theories have to be protected from refutation by auxiliary hypotheses. Cartwright (1983) makes a related point, arguing that the highly idealized theories of physics do not strictly speaking describe reality in any direct sense. But because of the explanatory value of these theories, it still makes sense to persist with them. Even some of our most successful scientific theories, such as Newton's laws of motion, are not necessarily supported or rejected by observation data, but only hold under some very specific and ideal conditions.

We see, then, that theory is not the underdog of data. Data that indicate anomalies from an accepted theory could be explained away by ad hoc hypotheses or simply be dismissed. This gives us the idea that data are not sacrosanct. Next, we will argue that data are already influenced by theory and that, without theory, we would even struggle to collect data.

11.5 Could There Be Raw Data?

There are good reasons to be sceptical of a purely empirically based science, and we have here presented some. There is, however, a further point to consider, namely the ideal of theory-independent or 'raw' data.

The ideal posed by Bacon is that our observations should be theory-neutral and not influenced by our expectations. This is to guarantee the objectivity of science; and the alternative does not seem very attractive. Any scientific publication must state conflict of interest for the authors—for instance, whether the research is industry-sponsored—exactly because we believe that personal and professional biases can affect scientific results. One might try to argue that the data are objective, but that the way in which we interpret them can be swayed, for example by financial interests or theoretical preferences. But can data ever be perfectly neutral?

To even get to the point where we can start collecting data, we need to make a number of non-empirical choices. Which methods should we use and which scientific tools are required? Whether we go for methods that are statistical, experimental, comparative, or purely observational is not something that itself can be determined by data. Another non-empirical aspect of science lies in the challenge of dealing with theoretically defined entities. Electrons, genes, photons, bacteria, and neurons can only be studied within the scientific framework of which they are a part. What, then, about even more abstract entities, such as history, culture, institutions, ideals, time, or causation? It seems that in order to do research involving any of these things, we must first define what we take them to be. None of these issues can be resolved based on empirical data.

Some philosophers argue that non-empirical assumptions shape science already from the outset, before we can even say that we made an observation. If we understand science in that way, data comes after theory rather than before. Experience is already cognitively mediated, one might think, as suggested by Hanson (1958: ch. 1). He says that we don't first experience data and afterwards have different interpretations about them. Rather, our background beliefs and knowledge directly affect what we can observe. Two people looking at the same thing could actually see differently, for instance, a scientist and a layperson looking into a microscope. To see something, we must first learn what to look for, which to Hanson is a theory-dependent matter. The layperson would lack the necessary theoretical knowledge and experience to see, for instance, the genes, proteins, blood cells, bacteria, or cancer that the scientist sees. To be able to see what the scientist sees, the layperson must first learn some science (Hanson 1958: 17).

If Tycho Brahe and Johannes Kepler were sat together on a hill watching the dawn, it is thus Hanson's claim that they would see different things because they have different theories that describe what they observe as different events. Kepler would see the Earth moving, while Tycho would see the Sun moving. The only sense in which they can be said to see the same thing is the retinal response of the light hitting the eye, but this, Hanson argues, is not seeing. An unconscious person would see nothing, even if the eyes are open and the Sun hits the retina; 'People, not their eyes, see. Cameras, and eye-balls, are blind' (Hanson 1958: 6). Observation is a conscious activity, not simply a matter of input and output. To see something as x involves theoretical and propositional knowledge about what being x entails; what x can do, or their dispositions:

Seeing an object x is to see that it may behave in the ways we know x's do behave: if the object's behaviour does not accord with what we expect of x's we may be blocked from seeing it as a straight-forward x any longer. Now we rarely see dolphin as fish, the earth as flat, the heavens as an inverted bowl or the sun as our satellite.   (Hanson 1958: 22)

That science should start from raw observation data is then a false ideal. As noted by Popper (1972), the mind is not an empty bucket that gets filled up when we passively receive sense impressions. Instead, it is more like a search light, actively picking out those elements that we take to be relevant or interesting. Without any theoretical knowledge, what we see would be literally meaningless. We could perhaps hear a noise or see some light or shapes. But even if we somehow managed to record these raw, uninterpreted, and unprocessed data, they would not be of the sort that could help us generate scientific theories. Instead, we would need theory to make sense of them.

11.6 Data or Theory? Would That It Were so Simple

Philosophy often deals in abstractions in attempts to find what is logically prior: as when Hume considers 'Adam' first looking at water and considering its causal powers, or Hobbes thinking of how man would behave in a state of nature. The question of what comes first, observation or theory, should be placed in this tradition. The question is based on the false premise that data and theory can be clearly separated. We should not, therefore, be tempted to say that we can have observation without theory, but nor should we think there could be theory without observation.

The relationship between data and theory is thus a complicated one. Hanson denies that we first see something and then interpret what we see. To him, data and theory are much more integrated. It would then be a mistake to think we can have one without the other. We always have both and they constantly challenge or reinforce each other throughout our ongoing scientific activity. We will be saying more in the rest of the book about how evidence and theory can be developed in unison. In Chapter 12, we consider further how data relate to causal theories, paving the way for an account of causal mechanisms. We should also try to understand why more was asked of the data than they could possibly deliver.

12

Are More Data Better?

12.1 If Only We Had More Data

We finished Chapter 11 by arguing that a theory has to be developed in conjunction with data as part of an ongoing process of investigation. Data and theory are developed together. This might still give the impression that more data are always better. It could be natural to think that the more data we have, then the more proportionally confident we should be in our theories. This might be, then, another norm of science and it would have clear application to the discovery of causal connections. The norm would tell us that the more data we have, the better position we are in to know the causal truths of what causes what. An associated idea would be that, ultimately, if we had a complete data set, one that included all the data, we would also know all the causal facts. What more could there be to know, it could be asked, in order to establish the facts of causation?

The acceptability of this idea depends, like much else that we find in the causal sciences, on our other ontological commitments: what we think are the most fundamental features of reality. This is a philosophical matter. Nevertheless, as we will now go on to show, there is an argument that in adopting the sorts of norm introduced above, science will have already taken a stance in this debate on these ontological issues. The norms in question suggest commitments that are favourable towards Humeanism and this may influence scientific practice, if only implicitly.

There are, for instance, two principles in Hume's philosophy suggesting that data generate causal truths. The first is his view that the world is just a collection of accidental and unrelated objects or events, where, in principle, anything could follow from anything. Lewis developed this idea into a metaphysical framework, called Humean Supervenience: 'the doctrine that all there is to the world is a vast mosaic of local matters of particular fact, just one little thing then another' (Lewis 1986b: ix). Within this 'Humean mosaic', there is nothing but the spatiotemporal arrangement of local properties. It is as if God just threw all the constituents of the world in the air and let them fall into a random arrangement, but in which we sometimes find patterns or regularities. Lewis used his Humean Supervenience thesis as a philosophical framework to account for causation, laws, counterfactual dependences, and a number of other things.

A second Humean principle supporting the pursuit of more data is the two-event model of causation; that causation is simply a matter of two events being constantly conjoined. We can see how this notion of causation follows neatly from the idea of the world as a mosaic of independent events. The way to establish causation between two events is to check whether they are of a type where one always follows the other.

An alternative to Humeanism is a commitment to causal singularism. On this view, causation is an entirely intrinsic matter, and happens in the concrete particular instances. Singularism is often found in philosophical theories that emphasize causal mechanisms, powers, and individual propensities. As we will go on to explain in Section 12.3, singularism challenges the norm that more data are always better for finding and understanding causation.

Evidence-based policy and practices, in contrast, seem in compliance with the norm, with an emphasis on the epistemic value of population studies and statistical models for finding and justifying causal evidence. It is stipulated that a randomized controlled trial (RCT) is better for establishing causation than a single case study, and that metastudies of RCTs provide better and more reliable evidence than individual RCTs. This illustrates a commitment to the norm that the more data, the more justified our causal hypotheses. To not fall victim of mistaking accidental correlations for causation, we must repeat studies and gather additional data until we eventually have a large enough sample. Only then can we make a qualified inference from the data set to the general population: that the intervention will work in general.

12.2 Humeanism: Treating Causal Inference as Inductive Inference

But is the norm correct or even appropriate when it comes to our understanding of causal theories? Would there be any sound basis for rejecting it? One reason would be if there are cases where the collection of larger quantities of data added little or nothing to our understanding of causal matters.

What might motivate the norm—that more data are always better—is treating the search for causal knowledge as if it were an inductive question; of inferring general causal connections from some finite set of observed cases. One common response to this inference, as suggested by the inductive method of Bacon (1620), is to observe as many instances as possible. This makes perfect sense on a particular view of how causal theories are developed.

Suppose we have a large bag of marble balls, the colours of which are unknown to us. If we want to know what colour they have, we can check a sample. If the first ten balls we check are all black, we might form a hypothesis that the marble balls are all black. But we might be wrong, since the next ball we check could be red or any other colour. In this scenario, it's attractive to think that the closer we get to the end of the bag, the closer we are to making a correct statement about the colour of the marble balls. The more data we have, the closer we get to the truth.

If causal knowledge were understood as analogous to the bag of marble balls, then more data would indeed get us closer to the causal truths. The causal truth would then simply be a generalization over all the particular instances, as in Hume's covering law view. Causation, on this view, happens in the concrete because there is a causal law that says so. But the causal law itself is nothing but a perfect correlation between two types of events:

A causes B, if and only if, a^1 is followed by b^1, a^2 is followed by $b^2 \ldots a^n$ is followed by b^n.

The bag of marble balls is an easy case to deal with, since it has a finite number of instances. Most cases of causation, however, would also involve future instances, so could in principle mean that we are dealing with an infinite number of as and bs. How, then, can we ever be confident about our causal inferences? If only there was a way to test every single instance, we could establish a constant conjunction of A and B and get a precise answer about our causal hypothesis. At least, that is the idea of how the problem of induction would go away.

It is easy to see why inductive inferences remain a bugbear for empiricism. As Hume argued at length, any conclusion about unobserved instances is empirically ungrounded and, according to positivism, at risk of being deemed unscientific. But this is where Lewis' Humean Supervenience thesis might be thought to have an advantage. On his account, the causal facts depend on the four-dimensional totality of non-causal facts or events (Lewis 1973a). If we knew the full mosaic of events throughout history, past, present, and future, then that would effectively fix the facts about what causes what. This is simply the omnitemporal version of Hume. Hume said that if you know what happens, and hence what is constantly conjoined with what, then that's all there is to what causes what. So Lewis is making the same move but considering facts from all times.

One might have some concerns about this as a solution to the problem of induction (not that Lewis was himself offering it as such a solution). First of all, not everyone might be willing to buy into the metaphysical extravagancies of Humean Supervenience. Second, even if one were willing, the account fails to provide a serviceable scientific method since no one could ever have access to future data. Hume himself saw that the need for complete data is a problem with his theory of causation and admitted that we, because of the problem of induction, can never have certainty for general causal knowledge. Popper followed Hume in this conclusion, but suggested we try to falsify our causal hypotheses and see if they survive.

It seems that we are left with two options. Either we continue to collect data until we have a complete data set, or we admit that all general causal claims are inductive, and therefore unscientific, even with a very large data set. Some might go for a third option and make a probabilistic claim instead, but as Hume showed (1739: I, iv, 1: 182–5), any finite number of instances of the outcome divided on an infinite number of possible instances results in zero probability of the outcome. We will return to probabilistic causal claims in Part VI.

If causal theories are based on inductive inferences, things are starting to look a bit bleak for the possibility of causal knowledge and justification. But are all causal theories basically formed by inductive inference? We argue no. There is a different way of understanding causation metaphysically which, if true, would undermine support for the norm of science we have been considering: that more data of the same type is always preferable for establishing causation. This view, causal singularism, is an alternative to Hume's generalist covering law view and offers room for epistemic optimism.

12.3 Dispositionalism: Causation Is Singular and Intrinsic

Let us start from the perspective that causation is an intrinsic matter, happening in the concrete, singular instance. What would be different? Take two events and consider whether they are causally connected. Hume did this with two billiard balls, where he observed that one moves towards a stationary second, they touch, and the second moves away. What would a singularist have to consider in order to conclude that the first ball caused the second to move? Would one need, as Humeanism suggests, to consider events occurring at other times and places—such as events in which similar balls are involved? Arguably not. Would it really matter what balls do on the other side of the world or what similar balls do in a thousand years' time? According to the singularist, all that matters, to know whether the first event caused the second, is what occurs between the two balls.

That causation is singular seems intuitive because it shows how it is natural to think of it as an intrinsic relation between the cause and effect. In considering whether *a* causes *b*, we would not normally think that this depends on things at other times and places. The Humean covering law model, and Lewis' four-dimensional version of Humeanism, effectively deny the intrinsicness of the causal relation and thus deny singularism. But with it, they also reject the existence of causal powers (for more details on the debate between Humeanism and powers metaphysics, applied to contemporary physics, see Esfeld 2010).

In contrast to the covering law model of causation, singularism is the view that there could be a single instance of causation; that is, a causal connection between two phenomena that occurs only once. It is then possible that there be an *a*, which causes some effect *b*, but that this type of causation is never repeated, so it was not part of some bigger general pattern. Does this seem far-fetched? The creation of the universe from the Big Bang could be one such unique case of causation. Should we have to deny this as a cause unless the same happened again many times? Hume said yes. We say no.

Contrary to the Humean need for a complete data set including future facts, singularism seems empirically sustainable. Some of our greatest scientific discoveries were made with very little data. Could one learn about causation by concentrating on one particular instance? Perhaps if one carried out one well-performed experiment, there would be little need to repeat it at other times and places (Cartwright 1999: ch. 3, but astonishingly anticipated by Hume himself, of all people: 1739: I, iii, 8, 104f).

It is hard to offer a direct, a priori argument for causal singularism. We have this intuitive argument but, like much else in philosophy, acceptance of the claim is dependent on it being part of a larger theory that is overall more compelling than rival theories. We have already offered some general considerations against the sorts of theory we reject (Part II) and in favour of the sort of theory of causation that we prefer (Part III) and more will be said throughout the course of this book. We will now say a bit more about our preferred theory.

Dispositionalism is an anti-Humean, singularist theory that emphasizes the causal role of properties. It is the metaphysical view that properties have real causal powers (see for instance Harré and Madden 1975, Cartwright 1989, Mumford 1998, 2004, Mellor 2000, Ellis 2001, Molnar 2003, Heil 2004, Martin 2008, Groff 2013). Our theory of causation belongs in this tradition, but in a particular version that we call causal dispositionalism (Mumford and Anjum 2011). On this theory, a cause is a property that disposes or tends toward its effect but, as argued in Chapter 8, it does so without guaranteeing the effect since all powers can be counteracted by other powers. This is what we mean when we say that causation involves tendencies rather than necessity (see Chapters 9 and 10). In the cases where causation does happen, where the cause succeeds in bringing about its effect, it is because individual dispositions or powers manifest themselves. It could be a very simple case of causation, such as when a glass breaks. The disposition is fragility and the manifestation is that it broke. Dispositions come in degrees. Something can be more or less fragile. A wine glass is much more fragile than a car windscreen, for instance. Whether a glass actually breaks when struck will therefore depend on how fragile it is and how hard it is struck, not on whether all other struck glasses also break. Note that this is disputed by some dispositionalists, such as Mellor, who defines a disposition as something that invariably manifests itself when stimulated in the suitable way: 'A glass that does not break when so [suitably] dropped is at that time not fragile' (Mellor 1971: 68). But, as argued in Chapters 5 and 6, what counts as 'suitable' in this context is entirely relative to the success of the outcome.

When causation happens—when a glass does break—it changes its properties, which is a common feature of causation. The glass goes from being smooth and fragile to being broken into many sharp pieces. The effect is typically produced by the interaction between different dispositional properties. The broken glass is the result of the meeting of the fragile glass, the hard hammer, and the strength of the impact between them. We thus take causation to be a complex matter, with more than one cause for each effect (see Chapter 7). As noticed by Martin (2008: 48), an effect is produced by what he called mutual manifestation partners. The fragile glass and the hard hammer, together with the impact, produce in conjunction the effect of a broken glass.

There is also a temporal aspect to causation, but not the one Hume proposed. On his view, causation is a relation between two separate events at different times. Instead, we take causation to be a single event or process, starting with the coming together of dispositional properties, and lasting only throughout their mutual interaction (see also Anjum and Mumford 2017a). What we call the effect is then typically a threshold case. Say one holds a wine glass, and grips it harder and harder until it breaks. The power inflicted on the glass thus accumulates until it reaches the threshold for the effect. Such causal processes often take time, since the threshold is not met instantaneously when the powers come together. Turning water into steam, for instance, involves gradual heating of the water until it starts to boil.

It should now be clear how singularism plays a central role on our theory. Causation happens in the concrete case and because of the properties that are involved. But this also means that causation could be a perfectly unique matter. A causal set-up could have a unique set of properties with a unique set of causal powers, which challenges another norm of science inspired by Humeanism: that it is a requirement for causation that it can be repeated. A singularist would deny this norm, at least as being universally applicable.

A singularist might therefore favour different scientific methods for dealing with causation. If we want to study and understand causation, the singularist would urge that we acknowledge complexity, tendencies, interactions, and the temporal aspect of the causal process. But then we need to identify which factors are involved, not only one single factor or trigger. We must also find out which qualities or dispositions these factors have, and how they interact with each other. The possibility of interference is crucial to consider, since for each additional factor the interaction might be different. In medicine, this could be a matter of life and death. Two people who seem similar in all relevant aspects might still be different in the ways in which they respond to a treatment. If one of them has an allergy, for instance, or uses a medication that could interact with the treatment, the effect of the treatment might be fatal to that person yet beneficial for the other. The effect is not the same for all, since each individual meets the treatment with their own unique set of causal factors: biological and others. Basing our causal knowledge in perfect correlations would then be an impossible task, for medicine at least, since no two people are exactly identical. The same can be said for causation in history, psychology, biology, and ecology. From a singularist perspective, we should expect that if we add the same causal factor to two different causal set-ups, the causal interactions will be different and thus also the effect.

Hume's covering law model suggests that we should collect more and more data of the same type from repeated instances elsewhere. In contrast, causal singularism would encourage us to collect more data about the singular instance. The type of data that is relevant for causal knowledge would then be richer in the qualitative aspect, but it would therefore also seem poorer from the quantitative perspective. There is thus a sense in which the norm that more data are better is valid also from a singularist point of view. But 'more data' must not in this connection be understood as more data of the same type. Rather, it should be more types of data from the causal context in question.

12.4 The Need for Causal Theories

We saw in Chapter 11 that the connection between data and theory is a complicated one. And what we there said about the necessity of theory applies to the specific case of causal theories, which are needed to explain a plausible connection between correlated variables. We have considered here whether the more facts we know, the closer we get to discovering the causal truths. Are more data always better? Only if we think of causal theories as inductive inferences, we argued, involving a general claim about all past, present, and future instances, do we seem justified in taking this as a norm of science.

There is a very different way of approaching this, however, if we instead think of causation as singular and intrinsic. Causal theories are often developed in the light of close examination of single cases rather than by observing repeated instances. And there is a good philosophical basis for such a view, because whether a causes b is arguably an entirely intrinsic matter between a and b, and the properties involved in the process of getting from a to b.

So how can we know if a caused b from a singularist perspective? Ontologically, it is sufficient for causation that a causal power manifested itself, either on its own or as a result of interaction with other powers. Epistemologically, though, we seem to need more than simply a being followed by b. It could still be an accidental case. What, then, could the singularist provide that the Humean cannot?

First, we would emphasize the epistemological primacy of causal theory over statistical data. Only if we know how and why the cause brings about the effect, or at least contributes to it, can we say that we have deep causal knowledge. This is a challenge to evidence-based policy, which downplays the role of theory and promotes big data as a way to establish causal truths. In this chapter we have argued that our differences lie in a philosophical disagreement about metaphysics and the nature of causation (see also Anjum 2016). As dispositionalists and singularists, we deny the idea that causal theories could be generated from a complete data set. With this, we also reject the positivist ideal that causal theories can, in principle even, be based on data alone. We also deny the Humean claim that causal theories are inductive inferences, to be approached as a bag of infinitely many marble balls. As long as we understand the causal processes involved, what happens elsewhere and at different times is ontologically, and perhaps also epistemologically, irrelevant. Because we already know how and why the nuclear bomb works, for example, there seems to be little need for a large-scale repetition of trials to confirm its effects.

We are not claiming yet to have provided a convincing argument in favour of qualitative approaches. That will need more development in what follows. But we have tried to cast some doubt on an approach that is purely statistical and quantitative. Over Chapters 13 and 14, we will offer more detail to our positive account of causal theory. We will also say something about exactly what type of knowledge would be relevant for understanding causation from a singularist perspective. Particular attention will be given to causal mechanisms, another problematic feature for the strict empiricist.

13

The Explanatory Power of Mechanisms

13.1 Quantitative and Qualitative Approaches

We have been looking at data-driven approaches to causation, and how data could be used to discover causal facts. In contrast, over this chapter and Chapter 14 we shift the focus on to a very different way of investigating causation, namely through the discovery of causal mechanisms that underlie any regularities that are to be found. Such mechanisms need not be narrowly physical, in the sense of being analysed entirely in the terms of physics, or fundamental physics. There could also be social, historical, and economic causal mechanisms at a relatively high level in nature. We are not dealing with the 'mechanical mechanisms' of seventeenth-century mechanistic philosophy, therefore, which concentrated on the movement of tiny corpuscles in accordance with mechanical laws (see Psillos 2011 and originally Boyle 1674). We opt for a broader notion of mechanism, in line with recent work (Glennan 1996, 2002, Machamer et al. 2000, Machamer 2004, Bechtel and Abrahamsen 2005, Bechtel 2011 and Craver and Darden 2013). In this contemporary setting, a mechanism is understood as a reasonably stable structure of objects or parts, with a particular organization, that allows them to play a function. To understand the mechanism of a cause is to understand how it gets to its effect, or how the effect is brought about by the cause. We can think of the cause and effect as the beginning and end of a process, for instance, and the mechanism as what comes in-between, linking them. However, we will have a number of qualifications to add, once we have presented our own definition.

The division between the two broad approaches of empirical investigation—statistical and mechanistic—matches two basic trends in scientific practice. There is no reason why scientists cannot be involved in activities of both kinds and, indeed, the position of methodological pluralism that we defend tells us that both are needed. The division is rather clear, however. Statistical approaches are focused on discovering the facts of *what occurs* and *how often* in the known world. Statistics cannot, of course, discover this infallibly but may record the proportion of cases in a supposed representative sample in which some variable can be discovered. The second approach is focused more on the *how* and *why* questions of science and this is where causal mechanisms and theory construction come to the fore.

This also reflects the distinction that is drawn between quantitative and qualitative approaches in research. To put this in basic metaphysical terms, quantitative research is concerned with the discovery of numeric values; that is, quantities. Such studies are typically performed via statistical, computational, or mathematical tools. Quantitative research can tell us that the average age at death in the UK is 81 and that 2 per cent of the Norwegian population is vegetarian, but also that the boiling point of mercury is 356.7°C, the speed of light 299,792,458 metres per second in a vacuum, and that the mean distance from Earth to the Sun is 149.6 million kilometres. Because all of these discoveries involve quantities, hence numbers, it means that quantitative research inherently relies on measurement in a broad sense. Ascertaining the proportion of vegetarians in a sample of the male population over sixteen is a form of measurement, for example, as is counting the number of plankton in a sample volume of water and discovering the length of a rod or its temperature. Hence, all these investigations can count as part of a quantitative approach in science. And because the research produces data that have numeric values, such research is also conducive to statistical and computational techniques, which means that we can produce detailed meta-analyses. With such analyses we can find correlations, excluding confounders.

Qualitative research does not produce data that have numeric values. It doesn't immediately, at least, even though at a further stage one might seek to put numbers to the data. Qualitative research concerns data that are not typically measured; that is, not being assigned a numeric value. Instead, qualitative research aims at a more in-depth understanding of meanings, beliefs, opinions, reasons, causes, and contexts of smaller and more focused samples, collected via inspection, interviews, discussions, stories, documents, participant observation, or field notes, for instance.

Making this distinction a firm one, between quantitative and qualitative facts, might require commitment to a metaphysical thesis that there are irreducibly qualitative properties, which some could yet deny. We can think of colours as being qualities, for example, as when something is red or blue. Others might have a belief that these qualities can be reduced to quantities, however, for example if redness is just a quantifiable wavelength of light (Armstrong 1997: 57–61). Another example would be when a hospital patient reports feeling better than the day before, which seems very hard to quantify in any natural or accurate way. Suppose someone reports feeling happy but is asked to say how happy on a scale of 1 to 10. If the answer is 10, because the patient is feeling very happy, then clearly it creates problems if, the following day, she is even happier. A bounded scale for happiness, which has an upper and lower limit, doesn't reflect the reality that happiness has no strict limit. And yet we need a bounded scale in order for a numeric attribution to carry any meaning and allow relative comparisons. If you quantify your happiness with a numeric value of 16, but cannot say out of how many, then you might not be telling us anything useful. Data that report a quality such as happiness might then be impossible to quantify in any meaningful way, though this is not to deny that one could gather statistics such as how many people report that they feel happy. The quality itself, however, seems inherently unmeasurable.

How a cause produces its effect falls on the side of qualitative research. The concern is not with what occurs or how often but, rather, with what are the mechanisms of the cause in question. Through what process did the cause produce the effect? Quantitative data could be more superficial information, but about many cases or instances, sometimes measuring just a few variables. One might say that quantitative data contain information of a few types but with many tokens, while qualitative data contain many types of information about just a few tokens. Qualitative data is likely to be richer and more detailed about one instance or a small number of instances. In the case of mechanistic knowledge, this will be primarily about what the components parts are in the process leading from cause to effect and how, why, and under which conditions they are related or arranged.

13.2 Causal Evidence

Confronted with such a distinction, between quantitative and qualitative approaches, and the types of empirical data they produce, one might wonder which is best suited for the discovery of causation between distinct variables. Speaking in pure epistemological terms, there are good arguments in favour of both. We will often come to accept that there is a causal connection of some kind, for instance, without knowing the mechanism involved. There seemed to be sufficient statistical evidence linking tobacco smoking with heart disease, to take one infamous case, even before a mechanism was known that could explain *how* smoking caused heart disease (Gillies 2011). Such knowledge of a causal connection might be partial and insecure, and to that extent uncertain. But it still might provide, under certain conditions, adequate grounds for an intervention such as issuing a warning against smoking.

This approach is acceptable in a number of sciences, boosted by the success of recent advances in epidemiology and evidence-based policy. Using statistical methods, researchers have found, for instance, that having trees and green areas in a city improves health perception and that green spaces seem to have a positive effect on air quality, mental health, blood pressure, physical activity, and recovery from surgery (see for instance Dye 2008 and Kardan et al. 2015). Although the mechanisms of these complex causal connections cannot be stated, and we are probably very far from being in a position to identify them, the evidence could still justify the decision to include more green spaces in urban areas. One might also observe that doing so is a relatively low-risk policy intervention in that no harm seems conceivable from introducing green spaces, even if it were to turn out that there is no genuine causal connection to health benefits.

The example suggests that there can be reasons to believe there is a causal connection based on quantitative evidence alone, and such a belief might, of course, be true. But there is also a parallel case to be made on the other side; that is, there could be good grounds to believe in a causal connection on qualitative grounds alone—without any recorded correlation—where those grounds concern mechanistic knowledge. We can give a simple example and a slightly more complicated one. The simple example would be knowledge of a seesaw mechanism consisting of a rigid plank attached to a central pivot that is secured to the ground. Knowledge of the mechanism might involve some hands-on experimentation to inspect the rigidity of the plank and its free movement on the pivot. However, without a need for repeated, recorded, and quantified tests, one could be in a good position to conclude that one end of the seesaw going down would cause the other end to go up. This is causal knowledge gained from inspection of the physical mechanism, which counts, therefore, as qualitative data.

A second example concerns the effects of an element, such as ununoctium, which is element number 118 on the periodic table. This element is so rare that it exists artificially only. Its name was just a placeholder, although it has recently been called oganesson. It was first synthesized in the laboratory in 2002 and only a few instances have ever existed because it is so unstable. Nevertheless, from knowledge of its chemical structure, it is possible to predict its various causal properties, such as that its natural state would be solid and that it would be able to bond with certain other elements to form predicted compounds, were it to persist long enough. Due to the fleeting existence of these atoms, there is no statistical data concerning any occurrences of these effects. They are based so far on theoretical mechanistic knowledge alone.

This second example tells us something too about the notions of technology and innovation, which are vital to science. Technology tends to progress when we come up with new things rather than merely record existing things; that is, events that have already happened. Without a reliance on precedent, innovation has to rest on another basis such as theoretical and mechanistic knowledge. If one considers the invention of the Archimedean screw, for instance, one can see how, without studying quantitative data concerning repetition of instances, it would have been possible to work out how the turning of an inclined screw within a tube would result in the lifting of water. This could have been planned and designed before any physical prototype was constructed. Bringing the point into the present day, one can see that similarly a designer of an information technology, such as a smart phone, would typically design both the software and hardware—planning all the requisite connections—prior to physical objects being built that instantiate those mechanisms.

In defence of repetition, however, we should probably concede that if a technology is subsequently sent to trial and is unable to produce the expected effect, then we should say the mechanism fails, after all, to be the cause we planned it to be. It is no good if the mechanism works only in theory. To be useful, it must produce its effect in reality as well. So perhaps statistical data, which uncover repetition and correlation, might have the final say as the way to test for the presence of a causal connection. Russo and Williamson (2007) suggest that science needs both to establish causation: probabilistic evidence, such as statistical correlations or evidence from RCTs, plus a mechanism.

13.3 Placing Causation Correctly

Although we favour the search for plural forms of evidence, we have a reservation about necessarily favouring quantitative over qualitative evidence, such as mechanistic knowledge. Quantitative evidence alone may fail to answer all the questions we have about causation, and thus fail to provide deep causal knowledge. Our reservation is primarily ontological, but one that feeds into a methodological concern. The reservation is ontological insofar as causation operates in a process-like way. It is not simply that we move immediately from the cause to the effect, with nothing in-between. Causation typically takes a path, which may be extended and through stages, and this is what qualitative research can describe as the mechanism. A mechanistic understanding of causation can thus reflect a deep knowledge of how it works in particular cases whereas quantitative evidence gives lots of relatively superficial data, such as that one thing tends to follow another or that one variable increases and decreases with another. Mechanisms can therefore hold more power than mere regularities. They could offer, for instance, a firm basis for counterfactual reasoning, which is useful when considering interventions.

We find, for instance, that married men live longer, statistically. But is this a good ground to change or guide our behaviour? Unless we can offer an explanation—a causal one—of why there is such a correlation, there is no reason for us to conclude that marriage somehow positively affects life expectancy in men. Many correlations are purely accidental, as researchers are aware, or due to other confounding factors. Even if married men live longer on a population level, we would not expect that signing a piece of paper could itself help prolong a life. Instead, we might find that there are certain aspects of the typical marital lifestyle that had a positive health effect: features such as companionship and economic benefits. Indeed, researchers have looked for and found factors that are causally relevant for life expectancy: stress, diet, lifestyle, loneliness, depression, which can be counteracted by some of the benefits marriage brings (Harvard Men's Health Watch 2010). When people are together they tend to take more care with their diets, for instance, and although the companionship that produces this tends to occur within marriage, it can also occur without it and with just as much health benefit. Applying the Causal Markov (Pearl 2000: 30) condition will show us this. Marriage can be shown to be irrelevant to lengthened life if those with companionship and stable finances live just as long whether they are married or not. Also those who are without companionship and stable finances can be found to live shorter lives even if they are married. Of course, one might need a prior causal hypothesis before one even thinks to look for the statistical evidence that could back up these further claims.

It follows that we should be cautious of any purely statistical evidence that is presented in a causal vocabulary. Without knowledge of a process or mechanism connecting two variables, one should be wary of findings that use terms such as 'increase', 'prevent', 'improve', 'reduce', 'promote', 'affect', 'enhance', 'counteract',

THE THEORY OF MECHANISMS

and so on, based on quantitative evidence alone (Cartwright 2007a: 19, 20). These verbs suggest that some factor or exposure is somehow responsible for bringing about a certain outcome. 'Outcome' is itself causally laden, suggesting an effect, which again means an effect of something (a cause). It might be a statistical fact that married men live longer but it could be false even to say that marriage 'increases' life expectancy.

Any causal theory that is lacking in a mechanistic aspect will be incomplete in some sense because it will have no account of how the effect is produced by the cause. To state the connection dramatically, you might be able to say that the cause *is* the mechanism, where it is the description of the path to the effect. But you cannot say that causation *is* the statistics or the randomized controlled trial, even if statistics sometimes provide good evidence that there is causation present. We might also be able to use quantitative techniques to discover a part of a mechanism, for that might be how a particular causal pathway is first detected in a previously incomplete theory.

To summarize thus far, if the aim of science is to offer theories, explanations, and predictions, causation is essential. This does not mean that we can do without correlation data. They can give good indications of where to search for causation. But our confidence in the existence of a causal connection will be greatly increased if we understand the causal processes involved too. Quantitative data might draw our attention to a possible causal connection but we then want to understand why and how something affects something else. And this is the stuff that scientific theories are made of: the what, the how, and the why.

13.4 The Theory of Mechanisms

The acceptance of mechanisms in causation is an important step but it does not end the matter. There are different ways in which mechanisms can be understood. Glennan introduces the idea as follows:

The term 'mechanism' is commonly used to refer to a variety of systems or processes that produce phenomena in virtue of the arrangement and interaction of a number of parts. The term was originally applied to products of human beings like watches or water wheels, but has, at least since the seventeenth century been used equally to describe systems (like cells) or processes (like those that produce sunspots) that are of natural origin. (Glennan 2009: 315)

As he then explains, there have been two strands developing this idea (Glennan 2009: 325). A mechanism can be understood as 'a network of interacting processes' (Railton 1978, Salmon 1984) or as systems: namely, 'organized collections of parts' (Machamer et al. 2000, Glennan 2002, Bechtel 2011).

Our own view of mechanisms differs from these two accounts though it is broad enough to accommodate what is appealing about each of them. As we have said, a mechanism explains how and why a cause produces its effect. On our account, it is an interaction of powers that constitutes the mechanism and that plays this explanatory role. This constitutes our definition: a mechanism is a complex organization of powers that are capable of producing a certain kind of effect through their interaction. Interacting powers typically issue in processes: the process leading from cause to effect. But a collection of parts can also count as a mechanism where those parts are empowered. Indeed, parts can only belong to a mechanism if they are powerful, which enables them to have a function. Our account of mechanisms is thus grounded in powers, hence we see no prospects for reducing powers to mechanisms. And because mechanisms are grounded in powers, our account differs from most other accounts in at least the following three ways. We say:

1. Mechanisms do not bring about their effects by necessity. Like any causal process, they can fail to reach their completion.
2. Mechanisms involve activity in the dispositional sense, which need not be occurrent activity; that is, it need not be a change at the macroscopic level revealed in events.
3. A mechanism does not necessarily involve a priority of the parts over the whole. It is possible to combine acceptance of causal mechanisms with holism and emergence.

We have already offered our reason for point 1, which is that causal processes tend toward their effects rather than bring them about with necessity (Chapter 10). This means that mechanisms are vulnerable to contextual interferers (Rocca et al. forthcoming). In Chapter 14 we will explain our commitment to the third point, that mechanisms need not be about the priority of the parts over the whole.

Here we can say something about the second point, which concerns the notion of activity. This matter is worth addressing because it is now standard to think of causal mechanisms in terms of active processes of multiple components that are organized in a certain way. We favour a dynamic view of nature in which change is the norm (Anjum and Mumford 2018b). If one considers the mechanism by which an aeroplane becomes airborne this is not just a matter of having physical, empowered components, such as the jet engines and wings, but also that the powers of these components must be exercised with a certain temporal organization. The engines must give the plane at least a minimum groundspeed before the flaps on the wings provide an upward thrust. To lift the wing flaps without speed would do next to nothing. Mechanisms are not just about the arrangements of components at a time, therefore, but typically will proceed through a temporally extended and ordered set of steps.

However, while causal mechanisms will normally involve activity, Humeans and dispositionalists might disagree over what constitutes such activity. It is possible to have a mechanism in which nothing is changing, at least at the macroscopic level of events. For example, we might have two magnets that are kept a stable distance apart by the positioning of a spring between them. The forces involved in this mechanism are perfectly counterbalanced, let us suppose, such that nothing is happening (see also Chapter 16). The case might be rejected by Illari and Williamson (2012; see also Illari and Russo 2014: 132), when they say that there is no mechanism without activity. However, a realist about powers is likely to think of this case as involving the action of the components even though the action produces no change. This is what is meant by Harré's notion of a generative mechanism (1972: 116) that, as Bhaskar (1975: 221–30) says, acts 'transfactually'. This means that its action does not produce the kind of events upon which the classic empiricists thought we should base empirical knowledge. There are no changes, nothing to record, but there is the action of powers that underlie, and indeed produce the stable arrangement of the components. We say that there can be mechanisms without change in this sense.

We are aware that an opponent might object that whenever a component power is acting, at least something must be occurring, even if it is at a microscopic level. For example, gravitons might be exchanged and electrons will be configured into magnetic moments. We need not deny this, indeed it confirms the idea of dynamism—change—being the default setting of the universe. But these underlying changes are still capable of creating changeless stability at a higher level of nature. At a macro level, a mechanism can be stable and changeless. One then might insist that since all macro-level phenomena are ultimately to be explained in terms of their components, what goes on at the lower level is all that matters. But this reductive assumption—that the parts govern the whole—is what we will challenge with our third point, and which is the topic of Chapter 14.

We have outlined the difference between quantitative and qualitative research in this chapter and explained how the evidence provided by these two approaches bears on the discovery of causes in science. Quantitative research can provide us with a lot of information of a certain kind: it can provide good evidence of what happens and how often, within a tested sample. With use of statistical analyses, quantitative approaches might help reveal where there is likely to be a causal connection. But, we then argued, a deeper knowledge is also preferred. Only when we have this qualitative understanding of the causal mechanisms that are involved will our confidence increase that we really have successfully identified a genuine causal connection. Like all knowledge, however, this remains fallible. We then argued for a distinctive account of causal mechanism. Mechanisms consist in an assemblage of organized components that are capable of providing a pathway in the form of a process that leads from the cause to the effect. Next we will dig deeper into the nature of these mechanisms.

14

Digging Deeper to Find the Real Causes?

14.1 Underlying Mechanisms?

There is an outstanding problem with the idea of causal mechanisms and mechanistic explanation. In seeking underlying mechanisms, a suggestion is that all causal explanation should be in terms of relatively lower-level phenomena, such as the arrangement of component parts.

The idea of explaining the behaviour of wholes in terms of their parts is both attractive and useful. Chemistry is a rich source of such examples. Consider the property of solubility possessed by substances such as sugar. We have an object, a sugar cube, which bears the property of solubility. We can think of solubility as a relatively high-level phenomenon. It belongs to some ordinary, everyday, macroscopic objects and occasionally we will see them dissolve in liquid. But chemistry tells us that there is a lower-level explanation for this observable phenomenon. In one such case, the relatively weak intermolecular forces in sucrose molecules have a stronger attraction to the H_2O molecule, when water is introduced, which effectively pulls the sucrose molecules apart. This gives us a lower-level mechanistic explanation of a higher-level causal process.

Similar explanations can be found elsewhere in science. Some illnesses can be explained with reference to bacteria, some organic development can be explained with reference to genes, some mental activity can be explained with reference to neurons, and so on. It would seem natural, then, that medicine, biology, and psychology all start to dig deeper to find the real causes underlying their science. And each explanatory success of this kind adds credence to the idea that there is a mechanism underlying every cause–effect relation. Perhaps all causal processes could be explained at a lower level, which suggests an asymmetric, one-way priority in causal explanation. Such a view suggests also a metaphysical thesis: that all determination is bottom-up. Parts always determine wholes, on this view, and that is why we explain higher-level phenomena in terms of relatively lower-level phenomena.

But not all adequate explanations of why something does what it does can be given solely in terms of its parts, we argue. We need to understand also how those parts interact and, in some cases, compose to make higher-level, holistic phenomena that are causally independent from the parts that produced them. The properties of wholes might rest upon a lower-level base that creates them but, nevertheless, those properties can be more than just the sum of the parts, in some special sense. We will call these emergent properties. This would mean that causal explanation—even mechanistic explanation—is not always to be given in terms of the parts of larger wholes. Instead, the most useful causal explanation of some phenomenon might be relatively high level, where the explanatory properties are emergent. However, we argue that there is no reason why we cannot reconcile the ideas of mechanism and emergence if we accept that: a) strong ontological emergence is compatible with it having a bottom-up explanation and b) mechanisms are not necessarily lower level in relation to what they explain.

14.2 Reductionism in Science

Reductionism is the philosophical thesis that all higher-level phenomena should ultimately be explained by lower-level phenomena (see Dupré 1993: part II, for different forms of reductionism). One can see how this thesis could be extrapolated from the putative successful reductions mentioned above. However, a finite number of cases, even if they really are successful reductive accounts of higher-level causal phenomena, would not justify the adoption of reductionism as a general thesis, applying to every case. That would seemingly be nothing more than an inductive inference from a few instances. It seems more, then, that there are philosophical commitments behind the thesis of reductionism: perhaps an idea that this is the way that causal explanation in terms of mechanisms ought to proceed. Perhaps there is something satisfying and complete about reductive explanation that would be lacking from an alternative view. There are reasons to resist this move, though, as we will argue here.

Some philosophers think that there is a lowest level in nature and that this is the level with which fundamental physics deals (for example, Oppenheim and Putnam 1991). Here we could find the simple and basic building blocks of reality: entities such as electrons, hadrons, quarks, and bosons. If that is the case, then all other sciences could be thought of as reducing ultimately to physics. One might surmise that the causal connections in sociology are in theory mechanistically explicable in terms of the psychology of the individuals within that society, that the psychology of individuals is to be explained in terms of neuroscience, that neuroscience is explicable through biology, biology through biochemistry, and, completing the chain of explanation, biochemistry is accounted for in terms of physics. If this is true, it is a very significant matter. It means that there is really only one causally efficacious level to which all others are reducible, namely fundamental physics. All other sciences would really be serving that. And, although we say that reductionism is a philosophical theory, there can be little doubt that it is a theory taken seriously within the sciences (Dupré 1993, 2001, Skaftnesmo 2009). There are a number of biological reductionist research programmes, for instance, such as evolutionary psychology (Dawkins 1976, Wright 1994), sociobiology (Wilson 1975), Darwinian medicine (Nesse and Williams 1996), and so on. Furthermore, neuroscience has a long tradition within psychology (Libet et al. 1983, Libet 1999, Libet et al. 1999), and has later moved into areas such as economics (Glimcher 2004), marketing (Hazeldine 2013), and moral judgement (Haidt 2001, 2007, Hauser 2006).

Although there are some successful reductions, and reductionism has proven progressive in many fields, such as understanding viruses in medicine, there are remaining problems with assuming that everything is reducible to lower levels of nature. In particular, we think that there are ontological and explanatory limitations on reductionism that require a more sophisticated and, at times, egalitarian view of the levels in nature. Most significantly, we argue that reductionism cannot account for the causal independence of certain higher-level phenomena that we should think of as emergent. There will be major methodological limitations on the sciences if we try to ignore these emergent phenomena. Such a view justifies the separation and independence of the so-called special sciences because, for instance, psychology is not ultimately reducible to fundamental physics. Explaining how this is so is not easy, however.

14.3 The Case for Emergence

In the first place, we will have to consider the issue of levels in nature. This is the question of whether nature really is stratified into distinct, hierarchically ordered, and discrete levels: is there a physical level, a chemical level, a biological level, and so on? What would be the basis for such a division? Is it simply that small things are lower level than big things? We will not dwell too much on this difficult subject, however, partly because both parties to the current debate—reductionists and emergentists—need some notion of level of nature in order to articulate their views. Reductionists think that higher-level phenomena are explained entirely by lower-level phenomena while emergentists think that some higher-level phenomena have causal independence from the lower levels. So levels in nature are not the bone of contention. Our own view is that a relatively innocuous distinction can be drawn between the higher and lower levels in terms of part–whole relations. If one kind of entity, such as an electron, can be a proper part of another entity, such as a table, then it is a relatively lower level than it. This would need refinement in order to constitute a full theory of levels.

With that working distinction, however, we can start to explain a case for emergence, and it is a case that we think to be perfectly scientifically acceptable. The basic idea is that there are some properties or causal powers that belong to wholes but do not belong to their parts. The cases that we call emergent are the ones where such new properties of wholes are a result of the causal interaction and transformation of parts, hence we call this the causal-transformative model of emergence (Anjum and Mumford 2017b). For other types of emergence, see Wilson (2016).

Properties of particulars we take to be causally powerful, which means that having a property must make a causal difference to something: a difference of what it is able to do or what it is able to have done to it. One could even say that properties are causal powers (Shoemaker 1980) but we don't need to settle that matter to give our account of emergence.

There are some relatively mundane cases where wholes have properties that their parts do not have; for example, an object could weigh 10kg though none of its proper parts do. But this does not count as emergence for us because the weight of the whole is just the aggregation of the weights of the parts. The weight of the whole is not a result of the transformation of the parts through their interaction.

In giving a causal account of how emergent phenomena arise, we risk the account being dismissed as mere horizontal emergence instead of vertical emergence (Gillett 2016). Horizontal emergence is simply when a new property of a particular appears at a later time, which was not present at an earlier time, such as when a baby is born with no hair but at a later time has a head of hair. Horizontal emergence is a diachronic relation concerning what is the case at two different times. Vertical emergence is the kind that goes up to higher-level phenomena, meaning that a new property emerges at a higher level that is not present at lower levels. Vertical emergence is supposed to be a synchronic relation; that is, the emergent phenomenon and its lower-level base exist at the same time. Typically, philosophers seek accounts of emergence in terms of constitution relations, where the parts constitute the whole, and constitution is a synchronic relation. However, this need not prevent us from offering a causal account, because we take it that an effect can exist at the same time as the cause producing it, such as when someone's arm holds up an apple or sugar dissolves while the solvent is present. The cause and effect existing simultaneously, does not entail that they must be instantaneous. They both can exist over a duration, because causes typically take time to complete their work. But, on our account, they exist for the same duration. The effect begins as soon as the cause is fully in place, which might be the joining of a number of different powers. And the effect ceases once the cause ceases to act; for example, the dissolving ceases to occur once all the sugar is dissolved. But where this happens, for each instant at which there is the cause, there is also the effect. Thus, the synchronicity requirement can be satisfied in a case where causes and effect exist at the same time and a causal account of vertical emergence is thus permissible.

An example will help illustrate our causal-transformative model of emergence. Water has a property of being able to put out fires. But its constituents, single atoms of oxygen and hydrogen, do not have this property. Indeed, single atoms of hydrogen and oxygen are highly unstable and reactive. They usually combine into molecules of hydrogen and oxygen, respectively. These two are gasses that would both fuel a fire, rather than extinguish it. Water, instead, is a rather stable component. The whole has not inherited the properties of its parts. Why not? Chemical theory explains to us how the properties of an atom are given by its structure, including how many vacant spaces there are on its outer shell of electrons. It is these vacant spaces that allow an atom to bond with various others, or not, in order to form molecules. Oxygen and hydrogen are able to bond with each other because of this. They effectively share electrons in their outer shell if there are two hydrogen atoms to one of oxygen. They become bonded through a chemical and causal process of ongoing attraction, the result of which is a stable molecule that has liquid form at room temperature. The stability of the molecule is in part because there are no longer vacant spaces in any of the outer shells of the constituents. They have, thus, been transformed by their interaction. They have become a whole—a unity—with its own set of properties and causal capacities distinct from those of its constituents; and specifically not the addition of the properties the parts would have when outside of the whole. The resulting whole is causally independent from the parts in that the whole has an ability to put out fires quite independently of the fact that its components on their own, prior to their interaction that made them a whole, would not. The components, rather, have the potential to form gasses that would instead fuel a fire.

Naturalistically inclined philosophers have been reluctant to accept emergence partly because it can sound unscientific. Some emergentists spoke as if emergent phenomena simply appeared out of nowhere. Strawson (2008: 65) accordingly posed an *objection from bruteness*, which means that an acceptable emergentism could not have the existence of the emergent phenomena be just a brute, inexplicable fact about the way the world is. But it can be seen that our version of emergentism does not have that feature at all. We say that the emergence that comes from our causal-transformative account qualifies as what is called strong ontological emergence even though in theory it can be explained entirely naturalistically how emergent phenomena come about.

This goes against a well-entrenched way of understanding emergence that we think has led to a dead end in the debate. It is standard to draw a distinction between weak or epistemic emergence and strong, ontological emergence (e.g. Chalmers 2006). Weak emergence is supposedly where the higher-level phenomena are surprising, unpredictable, and not explicable in terms of the lower-level phenomena. We currently do not know exactly how life emerged from lifeless constituents, for example. But it could still be claimed that life is only weakly emergent because if one knew the full facts of biochemistry, then one could know how life emerges from lifeless parts. Weak emergence is thus about our states of knowledge and ignorance.

Strong emergence, however, is usually taken to mean that the emergent phenomenon is unpredictable in principle, so that even if one did know every single fact about the lower-level phenomena and how the lower level worked, one would still be unable to predict or explain the higher-level phenomenon. But this, we think, is not the way to understand strong emergence and it has led some to look for it in the wrong place.

We have a number of objections to previous conceptions of strong emergence. First, it is far from clear what *unpredictable-in-principle* means. Time dilation close to the speed of light was once unpredictable in principle, given Newtonian mechanics, but this was simply because we did not yet have an adequate set of principles. Even if something is unpredictable in principle now, we cannot be sure that it never will be, so this is an unreliable criterion for emergence. Second, contrary to the intention, this is still defining emergence in epistemic terms, even if they are negative epistemic terms. The idea is supposed to be that we cannot know the higher-level phenomena even if we know everything about its lower-level base. And, as Kim (2006) complains, giving an account of something like emergence in terms of what is not the case is bad in itself, because it doesn't really identify something in common to all cases. One can compare this with how things can be not-red without having a colour in common.

We think that we offer, therefore, a positive account that overcomes the problems created in the traditional discussion. Emergentists should not look for phenomena that are inexplicable: that would be subject to Strawson's objection from bruteness. They should offer a positive characterization of what must happen, and how, for something to qualify as emergent. And this should be nothing to do with our states of knowledge or ignorance but, rather, be about what the world needs to be like.

We claim further advantages, one of which is that our account seems to fit with some of the standard alleged concepts of emergence, for which the term was first introduced. As we already mentioned, life is sometimes thought of as an emergent phenomenon. Whether it really is emergent will depend on the exact empirical details. But there is a good *prima facie* case for saying so. Organisms have some lifeless parts. The main ingredients in a human organism, for instance, are oxygen, carbon, hydrogen, nitrogen, calcium, and phosphorus, none of which are alive or organic when situated outside an organism. It seems plausible, however, that when these ingredients enter into certain arrangements, they interact in a self-sustaining and potentially self-reproducing way. All sorts of chemical bonds are created, forming DNA, muscle, bone, and so on. These parts are transformed by their interaction and constitute a whole that is alive. The parts can be said to be alive; but only insofar as they are parts of an organism that is alive.

Another candidate for our account is mind, which seems to emerge from mindless parts such as arrangements of neurons. A neuron cannot think. Arguably, even a brain cannot think. A mind is something a whole person has, rather than any of its parts, though one can see that those parts, being in the proper causal relationship, is a vital part of the explanation of where a conscious mind comes from. Sadly, we are probably very far from understanding the empirical detail of how this occurs. We claim, however, that if at some future point neuroscience were able to explain how consciousness arises, then it would not undermine its status as emergent. Indeed, we would take this as the explanation of how consciousness emerges from mindless parts. It would most likely vindicate emergentism, not undermine it.

Having a mind is very important for the persistence of some organisms. People, for instance, can decide to take food and water, to avoid danger, to sexually reproduce, where to go, and what to talk about. A number of these decisions affect the component parts of the organism. To emphasize the causal independence of the mental, it seems more plausible that a minded person is carrying around her molecules than that the molecules are carrying around her.

14.4 Holism and Demergence

This allows us to return to the issue of mechanisms and why we should accept higher-level mechanisms that are not to be explained automatically in terms of their parts. Rather, there are some cases where the nature and persistence of the parts can be explained only by holistic mechanisms at a higher level.

Although emergent properties are created by the causal interaction of their parts, and are thus made from the bottom up, the causal independence of the emergent properties means that they are subsequently able to affect their parts top-down. We can call this demergence, which means that there are some properties possessed by the parts only because they are part of an emergent whole, which has caused them. An individual person, for example, can have a property of being a language speaker. Language is typically thought of as high level and emergent, because it is, at least partly, a norm-governed social phenomenon (Wittgenstein 1953: §243, §§256ff). But once someone is part of a language-using community, they are also able to speak and understand the language, as part of that community. Being a language speaker is the demergent effect, therefore, of a top-down change, caused by emergent linguistic phenomena.

Similarly, there are mechanisms in economics that can explain why inflation increases with money supply, why increases in price decrease demand, or even why certain economic systems require selfish behaviour among those subject to it. We cannot possibly reduce these causal mechanisms to the actions of molecules, indeed, not even to the actions of individual people. Economics is an emergent social phenomenon, essentially concerning how groups of individual people interrelate. Even though societies are composed of individuals, they also have a causal independence from them, which means that socioeconomic history, with its own powers, is not bound by the powers of the individual members of society. Indeed, economics can cause changes to those individuals, such as if someone suffers poor health because of poverty.

Because our account is a causal-transformative one, and causes tend toward their effects, rather than necessitate them, we cannot even say that higher-level phenomena supervene on lower-level phenomena. Supervenience is the technical term for a relation that can hold between two types of property. Property type B supervenes on property type A just in case if two particulars are alike in all their A-type properties, then they will be alike in all their B-type properties. On a dispositional account of causation, however, two identical causes might tend toward the same effect but, as there is no necessity of their effects, one or both might fail to produce it. Emergent phenomena, although often produced by certain initial arrangements of lower-level phenomena—what we can call their causal base—do not therefore strictly supervene on that base.

To allow irreducibly high-level mechanisms is, admittedly, a radical suggestion, which might seem in conflict with a core notion of mechanisms. Psillos (2011: 772) says of seventeenth-century mechanistic philosophy, for instance, that 'This priority of the parts over the whole—and in particular, the view that the behaviour of the whole is determined by the behaviour of the parts—is the distinctive feature of this broad account of mechanism' and, of mechanistic explanation generally, that 'a mechanical explanation is a kind of de-compositional explanation: an explanation of the whole in terms of the parts, their properties and their interaction' (Psillos 2011: 780).

Our account does not fit that. But the kind of irreducibly high-level causal explanation we have defended does meet much else of what Glennan (1996), for instance, or Machamer et al. (2000), would regard as a mechanism. It is still a causal explanation in terms of certain arrangements of entities and their properties that link a cause to an effect. What we do is leave open a possibility that explanation is found in relatively higher-level conditions.

Instead of suggesting a reductionist account of the world, then, we acknowledge different levels in nature but allow them more equality, consistent with Dupré's (1993: 97) egalitarian conception of levels in nature. This also has some support from Schaffer when he says that: 'Mesons, molecules, minds and mountains are in every sense ontologically equal' (Schaffer 2003: 512, 513). There are, of course, many cases where the parts have priority over the whole. But we have also made a case for wholes having priority over the parts, under some circumstances. We cannot assume, therefore, that the lower level always does the 'real causal work' and, in that sense, we cannot assume a strictly ordered, bottom-up hierarchy in the world's structure.

PART V

Linking Causes to Effects

15

Making a Difference

15.1 The Causal Link

In Part V, we will be looking at various issues concerning how causes link to their effects. This question is pertinent even if, for instance, one accepts that a notion of causal mechanism is useful. We saw that a mechanism consisted in an arrangement of components that played the function of taking the cause to the effect through a process, which could consist in a series of sub-processes in those cases where a mechanism was complex. The mechanisms could be lower level, where the behaviour of a whole is explained in terms of its parts, or higher level, where the behaviour of the parts could be explained in terms of them belonging to a larger whole.

Yet it might be thought that there still is a deeper question of what links cause and effect, even if we consider processes and sub-processes. One might wonder why, for instance, A causes B instead of causing C. One might wonder what the causal link is in individual cases: why a particular mechanism or sub-mechanism is for a particular end result. What justifies saying that an event was causally dependent on another? There can also be questions about longer links in causation: can they form an extended chain and are there any restrictions on the sorts of things that can be causes and effects? In particular, must causes and effects be 'happenings' or can they be 'non-happenings'? All these matters will be addressed in this part of the book.

The chapters here are thus loosely arranged around the theme of linking causes to effects though they can also be read in isolation, in that each is concerned with one of the standard and frequently discussed topics in the philosophy of causation. We start with whether it is the notion of difference-making that links causes and effects; and, in particular, whether causes are nothing more nor less than difference-makers.

15.2 A Methodological Assumption

The idea that causes make a difference is compelling. The effect would not have occurred without the cause, one might say. John Stuart Mill (1843: III, viii) spoke in general terms of a *method of difference*. If causes make a difference then it suggests to us a method of discovering them: check to see when the introduction of a factor makes a difference. It is easy to argue that this notion of difference-making motivates key methods in the causal sciences.

There are comparative studies, for instance, the sophisticated version of which is the method of randomized controlled trial (RCT). What justifies this as a method of identifying causes is that it reveals in a rigorous and scientific way the difference made by an intervention. The standard approach to RCT is, first, to divide a large number of subjects into two groups completely at random so as to have two groups that are as alike as possible; hence, they should have similar numbers of men and women, similar numbers of alcoholics, diabetics, optimists, manual workers, and so on. Second, one group then gets the trial intervention, such as a drug, while another group, the control group, gets nothing or merely a placebo. All those involved in the trial might have some condition that they would like to see resolved. Third, one records the outcomes for both groups and then analyses the data. Specifically, one would look to see whether the condition in question had a subsequent lower incidence in the treatment group that received the trial drug than the placebo group.

Suppose that the recovery rate from the condition was virtually the same in both the treatment group and the placebo group. Then one could conclude that the trialled intervention made no difference. There would be no point prescribing it in future, therefore, because there is no evidence that it will work. Suppose, on the other hand, that those receiving the drug recovered from the condition at a far greater rate than in the placebo group. That is when claims will be made that the drug can cause recovery. Such a causal claim is thus premised on the idea that the drug made a difference; and it's a difference revealed by comparing an intervention with a non-intervention.

RCTs make this comparison in large groups, but not all comparative methods do. In experimentation, for instance, one attempts to change one variable to see whether any other changes with it (Gower 1997: ch. 2). One might intervene on one test case, as in engineering. Perhaps one increases a variable for a single component, such as strengthening a rod. Does that make the component more efficient or reliable in fulfilling the purpose required of it? In veterinary science, perhaps one attempts an intervention on a single animal, for example, that seems to be having trouble walking. Surely the test of the intervention is whether it made any difference; preferably for the better. Or consider particle physics conducted in a large laboratory. An experiment introduces some change, such as smashing a hadron into another particle, and then the experimenter looks for what, if any, difference that collision made to the particle.

Comparative methods are useful, then, perhaps not because comparison is the end in itself—the thing that we want to identify from scientific practice—but because comparison can reveal whether differences were made through an intervention. What would justify this as a methodology would be the philosophical idea that causes make a difference; hence, if one can discover the difference-makers, then one has discovered the causes.

This idea is attractive, of course. Consider your own actions. If nothing that you did ever made a difference to anything, then you would feel causally impotent. You would feel you were not properly engaged with the world—perhaps not even a part of it at all—and thus not capable of acting. Rather, we know that we are able to act upon the world precisely because our interventions lead to changes in other things. You can call out for someone who hears you; you can press a key on your keyboard and the document is updated on your screen; you can take a drug and your headache stops; and you can put pressure on a wound to stop it bleeding. Arguably, we know we are causally active in the world exactly because we make a difference to it.

However, the question of whether this reveals the essence of causation is a more complicated one. Because of this, the connection between difference-making and our scientific methods will need to be cautious and refined. We will show how the connection between difference-makers and causes is not an exact, one-to-one match. In particular, there are some causes that are not difference-makers; and there are some difference-makers that are not causes. This complicates the picture because it means that if we identify a difference-maker, we have not necessarily identified a cause, and vice versa. Nevertheless, difference-making is a good symptom of the presence of causation, we suggest. Causes tend to make a difference; it's just that they don't always.

15.3 Causation without Difference-Making

Reflection shows us that we can have causation without difference-making, which means that we cannot take a lack of difference-making as a sure sign of a lack of causation. This possibility arises because a cause can operate in a context in which there are back-up mechanisms involved. A cause might be producing an effect but, were it to cease or fail to do so, another mechanism could take its place. In that situation, it might look empirically as if the first cause was not doing anything when in fact it was. But technically it was not making a difference because the effect could continue even were it to stop. It may be that in many cases of causation, the effect would indeed stop, were the cause to cease acting, but there are at least some examples where this is not the case.

Philosophers use the term overdetermination to describe cases like this (Lewis 1973b), though the term was used before that in psychoanalysis (Freud 1899). Overdetermination in the technical sense occurs where there are at least two separate causes in operation and each individually is capable of bringing about the same effect. This is also thought of as a variety of causal redundancy insofar as the effect could have occurred even if one (or more) of the causes had been absent, as long as at least one remains present. There are simple examples. In a firing squad, ten soldiers might shoot the victim, who dies. One bullet would have been enough, let us suppose, in which case nine of the bullets were unnecessary for the death. They were causally redundant in that the death could still have happened without them. But given that the victim took all ten bullets, his death was overdetermined. In another case, a traveller walking through the desert has two enemies. One puts poison in her water bottle. The other, without knowledge of the first, drills a small hole in the bottle so that it runs dry. The traveller is later found dead. There are less gruesome examples.

A wooden cross stood in front of a light can cast a shadow. But you could also position a smaller cross with the same shape closer to the light. The shadow is then overdetermined in that you could remove one of the crosses and the shadow stays there, unchanged.

These are cases where the same effect can occur if one, but not all, of the causes of an effect are removed. It seems that the cause that is removed was producing the effect, in the sense that it could have done so alone, and yet it made no difference when it was removed. If A and B overdetermine an effect E, then E would still have occurred if A continued and B ceased but also if A ceased and B continued. Both A and B, therefore, are individually adequate for E but also individually redundant for it in this context.

This distinguishes overdetermination from simple causal complexity. It is perfectly ordinary for an effect to require multiple different causal factors. Human conception requires a contribution from both a sperm and an egg, for instance; but these do not overdetermine the conception. Neither is redundant because neither is adequate alone. We can also distinguish a different case of causal redundancy where there are not multiple causes in operation but just one. However, this one cause has more than enough to produce its effect, which therefore could still have occurred with less. A table thrown through a window is enough to smash it but perhaps the same effect could have occurred by throwing a chair through, which would have had less force. The force of the impact from the table was indeed the cause of the window's breaking but, given that it could still have broken with less, then some part of that cause was redundant with respect to that breaking.

Now some of these cases might seem to be mere philosophers' fancies. However, we can point to biology as a case where overdetermination actually occurs; and for good reasons too. There is a clear survival advantage in having back-up mechanisms for any biological activity that is essential for life. Humans can breathe through their noses and their mouths, for example. If there were only one method of breathing, the risk of death would be greatly increased, for instance, if something got stuck in the nose. Back-ups can be more elaborate than that. In a well-known genetic experiment, drosophila had the genes removed that were known to be producing their eyes (Webster and Goodwin 1996: 87). Surprisingly, eyes returned to the flies after just a few generations. It was not that the removed gene had returned. Rather, other genes had taken on the role, supposedly because those genes were serving a need of the organism (hence a case of demergent, downward causation; see Chapter 14). It is easy to see, therefore, why there is an evolutionary advantage in having this kind of causal redundancy built into an organism. One can also see how it would make sense to deliberately build redundancy into a technology so that essential functions are backed up, such as the safety mechanisms of an aeroplane.

There is an ongoing discussion of overdetermination in the philosophy of causation (see Collins et al. 2004). Those who prefer difference-making accounts of causation will often deny that overdetermination ever really happens. Our view is that the chief motivation for denying the possibility of overdetermination, which otherwise seems plausible, is merely to salvage a theory that causation is difference-making. And where overdetermination does occur, it still may be perfectly scientifically scrutable. In all the examples given, it is perfectly possible to discover what all the separate overdeterminers are, such as the two crosses that cast the same shadow. The bearing of overdetermination on questions of scientific method, then, is that it blocks the simple inference from cause to difference-making. By *modus tollens*, then, it blocks also the inference from lack of a difference-maker to lack of a cause.

15.4 Difference-Making without Causation

For separate reasons, we should resist the inference the other way too. When we have discovered a difference-maker, we cannot automatically assume that we have identified a cause. Indeed, by our lights, we would not expect even that most difference-makers are causes (whereas it may be that most causes are difference-makers).

There is another reason why the notions of difference-maker and cause come apart. This is where we have a necessary condition for something to occur but where it is not really a causal contributor to it. For example, having a liver is a necessary condition for having cirrhosis. No one could have cirrhosis unless they had a liver; therefore having a liver is a difference-maker with respect to that disease. But it would seem wrong to say that the liver caused the disease even though it is a necessary condition for having it.

The argument against all necessary conditions being causes is a simple *reductio ad absurdum*. A necessary condition for something, E, is simply a condition without which E would not occur: a *sine qua non*. But for every event in the world's history, there are countless necessary conditions for them, some very distant, and which have no realistic claim to being causes. According to a widely accepted view, the Big Bang is a necessary condition for everything that came after it. But it would be explanatorily redundant to call it a cause of, for instance, World War II, the Great Depression, or the result of the UK's EU referendum. Without the Big Bang, there would not have been a vote to leave the EU in the referendum, but how could there be any meaningful sense in which the Big Bang was the cause of that result? Being a necessary condition of something tells us nothing about how that event is brought about. We can see this when we consider that the Big Bang would also have been a necessary condition of getting a vote to remain in the EU, which didn't happen. This shows that we need more than a necessary condition of something for it to be caused—because there are necessary conditions for many things that don't happen to get caused.

Of course, this leaves the question of how a cause differs from a *sine qua non*. We will answer that question in Chapter 16. All we need to claim here is that necessary conditions are not the same as causes. Examples are enough to illustrate this. Having a liver does not cause someone to get cirrhosis, nor does that person's birth. Similarly, her birth is not the cause of everything she subsequently does in life. Being a necessary condition of life itself, her birth is explanatorily redundant with respect to what she then does with that life. If we want to know the cause of her studying physics at university, for instance, we will want something more specific, such as that she had an inspiring physics teacher at school.

We have now seen how there can be causes without difference-making and difference-making without causes. This rules out an automatic inference from one to the other. Nevertheless, it is still the case that there is some, less straightforward, connection between difference-making and causation; and this is a connection that we can still exploit in our methods for the discovery of causes.

15.5 The Contrastive Approach

We started the chapter with a consideration of comparative studies. We can use such studies to identify what follows from a difference in exposure to a certain factor. This tells us something useful about both scientific method and causal explanation. Causal explanation sometimes works through drawing a contrast. If you want to know why Anne studied physics, a contrastive explanation is a good explanation, as opposed to citing a necessary condition. She studied physics instead of another subject because she enjoyed it at school more than anything else. And we have already seen how it helps to discover causation if there is a contrast between cases in which one variable is present and another in which it is absent (see, for instance, van Fraassen's 1980 account of 'why' questions and also Cross 1991).

There is, then, hope for a contrastive theory of causation (see Schaffer 2005). However, we argue that such a theory should be kept in its proper epistemic place and not treated as an ontological theory about the nature of causation itself. Contrastive relations—such as when we say that A, rather than B, caused E, or that A caused E rather than F—might be useful explanatorily and epistemically, but they have limitations because they do not map the nature of causation itself. As we saw in Chapter 13, there are reasons for treating causation itself as a non-relational matter, by which we mean one internal to the cause and effect, C and E. It might help us to understand C as a cause of E by contrasting it with something else, such as not-E. But the causing of E by C is not dependent on any such relation to not-E.

We can illustrate this point with reference to an argument.[1] In epidemiology, in order to claim that a trial drug is effective, we need to compare a recovery rate in a treatment group with the recovery rate in a placebo group. Without the latter, the recovery rate in the treatment group is deemed to tell us nothing about whether the trial drug did any work. If recovery rate is higher in the treatment group than the placebo group, then we can say that the drug worked. It made a difference. However, the existence of the placebo group may be a useful methodological approach to

---

[1] This argument was first presented to us by Roger Kerry, but is also treated briefly in Kerry et al. (2012).

discovering whether the intervention worked but it is in no way constitutive of the intervention having worked. Suppose 50 per cent in the treatment group recover from an ailment while only 30 per cent recover in the placebo group. We might assume from this that a number of people in the treatment group got better because of the drug: around 20 per cent of them. However, doing so had nothing to do with anything that occurred in the placebo group, nor even that there was a placebo group.

What if, due to an administrative error, no placebo group was set up in the first place and the trial was run on the treatment group alone? One has to assume that 50 per cent in that group would have recovered whether there was a placebo group or not; and 20 per cent of those who recovered were caused to do so specifically because of the drug. The point is that the drug doing its work on those in the treatment group is unaffected by anything going on elsewhere with other people. Of course, we might not know this without running a proper trial, but that concerns the issue of our causal knowledge, not the nature of causation itself. Thus, we are happy with contrastive approaches methodologically but not that they be taken as constituting causation itself; for instance, if someone claimed that C causing E consisted in nothing more than E following from C but not from not-C.

Why, though, is it so important to maintain this sharp distinction between epistemology and ontology? As we argued in Chapter 3, causation is the thing in the world that we are seeking with our scientific practices. If we have the nature of causation wrong, then this will mislead our causal judgements. We may have a method that succeeds in finding some cases of causation. But if, on that basis, we identify causation with success under that method, then there could be other cases in which we get the wrong results. In relation to the current discussion, there are at least some cases where causation and difference-making come apart. If we identify a cause with a difference-maker, then our causal judgements will fall into error in some cases. Our interventions could then fail in their intentions.

15.6 Difference-Making as a Symptom

What can we then do? How can we explain the relationship between the causal information we gain through comparative techniques and causal theories?

Causes tend to make a difference. We saw in Chapters 9 and 10 what is meant by a tendency. A tendency is something that is disposed to happen where it is more than a pure contingency that it happens but less than a matter of necessity. Hence, a tendency might be revealed in a raised incidence of some outcome: a tendential regularity that is less than a constant conjunction. We can make use of this idea here. Some causes make a difference. Perhaps most of them do. But not all of them do. We cannot say, then, that it is a necessary condition of something being a cause that it makes a difference. And we cannot, therefore, make a causal claim rest always on any such condition being satisfied. It should also be clear why causes tend to make a difference and why occasionally they don't. Causes produce effects and in many cases those effects would not have happened without those causes. In such cases, they make a difference. But we have also seen why this is not universal, for instance if there are overdeterminers present, such as back-up mechanisms.

However, the fact that causes tend to make a difference is enough for us to exploit in our scientific practices. The tendency allows us to treat difference-making as a sign or symptom of causation. A symptom is something that can be taken as indicative of something else (typically because it is a cause of it). For example, eczema is regarded by some as a symptom of stress. Now there can be cases of eczema that are nothing to do with stress, and of course there can be stress without eczema. But there are still rational grounds for treating eczema as a symptom of stress because stress has a tendency to produce it.

This is the analogy that we wish to exploit at a higher level of abstraction. Difference-making can be taken as symptom of causation because causes tend to make a difference. There can be one without the other but the fact that one tends to follow the other can be a rational basis for taking it as indicative, albeit fallibly indicative, of a causal connection. Stress causes eczema. Do we also want to say that causation *causes* difference-making? That looks like a bit of a metaphysical double vision. But we can certainly say that it is because there is a causal connection that, in many cases, a difference is made. And we can even say that it is the fact that A causes B that makes a difference with respect to B, given A. Thus, saying that a difference is caused is enough of a reason for taking differences as signs of causes.

We will be seeing that there are a number of other phenomena that can similarly be taken as symptoms of causation but, before we come to those, we will pursue the matter of the causal connection further.

16

Making Nothing Happen

16.1 Change and Stability

Not all causation is causation of change. Some of it is causation of stability; that is, non-change. Many theories of causation fail to represent such causings. Causation is often understood to be a relation between two events or occurrences, such as the pushing of a button and the firing of an electron. This could reasonably be called the orthodox view. If we think of the issue of the causal link, then the topic of this present chapter is more about what is linked rather than what is the link. And in standard ways of understanding causation philosophically, the things that are linked—the relata of the causal relation—are events.

What, then, of absences of events? There are some plausible cases where either the effect or the cause are absences of events: cases where nothing happens. There are even plausible cases where both the cause and the effect are absences of events. For an empiricist, this is a problem. Only in cases where one has observable events, could one say that there is causation.

The metaphysics of absences concerns the division between being and non-being: between existence and non-existence. Surely there can be no more fundamental division than that. It is then hard to explain this division in terms of something else. However, we will find that progress can be made on the issue of absences in causation while remaining solidly empirically grounded. A sensible position is possible that avoids obscurantism.

The position we advance is as follows. There is no reason why non-events cannot be effects. The causing of stability, we argue, is an important case of effects, despite its neglect philosophically. Much that occurs in the natural world is dependent on the existence of stability, including cases where it emerges out of change. It follows that the theories of causation are too narrow in which the relata of causal relations are restricted to events. Those theories will need amending or rejecting. Similarly, a cause can be a non-event. It could be, rather, a standing condition, property, or power. It may have changed at some point but could be changeless while causing.

We also claim, however, that a cause cannot be the absence of anything at all. Some have argued that there are cases like this: of causation by absence. But an absence, where it really is nothing, has no causal power, we say, so some other explanation must be given of why these look like cases of causation.

16.2 Stability in Action

We concentrate first on cases where an effect is a non-happening. Some would call this causation *of* absence (we will come to causation *by* absence later). What we mean here is specifically the absence of an event. It may be simpler to think of this as a case where the effect is that nothing happens: nothing changes.

It is easy to see how this type of case is important. Stability can be just as important as change, as an effect. An engineer who builds a bridge wants it to be a stable structure that, in most respects, is unchanging. The structure should be unchanging even though things happen to it. It can remain in place through a storm, for instance, and when people and vehicles stand upon it. There is a technical design that has enabled the achievement of this effect. Stresses have to be calculated and distributed throughout the structure. This gives us our central kind of case of an effect where nothing happens and it is easy to see how any number of such examples could be generated. We can call this kind of effect, where nothing happens, changelessness. And if events are essentially occurrences or changes, following Lombard (1986), then it is clear that any such changeless effect is not an event. Within the empiricist tradition, events are essentially occurrences.

It might be noted that some changes occur to the bridge even when it stays in the same place. Perhaps it undergoes some movement, such as when it expands in heat. Is anything, after all, completely static in this universe? What the example shows, however, is that stability in some respects is consistent with changes in other respects, which is important within biology in particular.

We can understand *stability-through-change* first with an astronomical example. This will be a case of equilibrium between counterbalancing forces; namely, the stable orbit kept by the Earth around the Sun, at a mean 93 million miles. Notably, this is a stability in respect of relative distance that could only be achieved through change, given that we are dealing with unsuspended astronomical bodies. What we have is a gravitational attraction between the two bodies that generates a centripetal force but also the relative motion of the Earth in its orbit that generates a centrifugal force. Fortunately, these are in perfect balance. Were the centripetal force stronger, our orbit would decay and we would crash into the Sun. Were the centrifugal force stronger, we would fly out of orbit. Even though there is movement, therefore, in respect of the relative distance between these two bodies, the effect is an unchanging stability. The distance stays the same even though other things change.

The reason this kind of stability-through-change is important in biology is that organisms must always be in motion (Anjum and Mumford 2018a). If a living thing ceases completely to move, it dies.[1] Animation must be constant. And yet, through this constant change, there are many features that have to be kept stable, or at least within a stable range, such as body temperature, heart rate, insulin levels, and oxygen

---

[1] Our thanks to John Dupré.

levels in the blood. How, then, is such stability achieved when change is required at all times? As in the case of the stable orbit, a biological 'solution' to this problem is to have ongoing processes that counterbalance. It then makes sense to think of these processes not as working against each other but as working *with* each other. Centripetal and centrifugal forces might be pulling in opposite directions and in that sense be competing. But in the case of organisms, it is for the health and survival of the living entity that this balance is achieved.

An old example illustrates the idea. This is George Molnar's (2003: 195) case of the horse-drawn barge. If just one horse pulls the barge, located on one side of the canal, it would drag the barge on to the canal wall. The solution is to have another horse on the other side of the canal also pulling. The two horses jointly pull the barge safely down the centre of the canal. These opposite forces are then working with each other to produce the desired result. It is teamwork. In more abstract terms, one mechanism of this kind, working alone, could not have produced the effect, so the problem is solved by having two mechanisms that are able to produce the effect through their joint and counterbalanced action.

In biology, then, we find complex homeostatic mechanisms that maintain stability through ongoing change. An easy such mechanism to understand is the maintaining of body temperature in humans where the body will sweat if too hot and shiver if too cold. The first cools the body down; the second warms it up. For the most part, these mechanisms ensure the body stays within a narrow range, either side of 37°C. Another example is the role of insulin in the body's use of sugar. Functioning normally, insulin ensures that blood sugar levels stay within a range. The pancreas releases insulin when blood sugar is too high (hyperglycaemia) and stores glucose in the liver so that it can be released later if blood sugar is too low (hypoglycaemia).

Stability and changelessness have a significant place in science. We often want to cause things to happen but equally there are cases where we want nothing to happen, for some situation or arrangement to remain the same. If we have a theory of causation that cannot account for it, or at least not very naturally, then it will not count as a good theory of causation. It seems an overly narrow theory of causation, therefore, that is restricted just to events.

16.3 Generative Mechanisms

We need to understand better what happens (or, rather, doesn't) when an effect is a lack of a change. The theory of causation should then take it into account and so should our methods in finding causes. Some philosophical background will be necessary in order to do this and it will involve a return to the idea of a generative mechanism, which we encountered in Chapter 13.

Hume's empiricist approach to causation has already been discussed (Chapter 5), mainly because of his view that we believe there to be a causal connection when we see a repetition or constant conjunction of events. Understanding the relata of causal relations to be events might seem perfectly innocuous and it will fit a lot of common-sense examples. Lighting the stick of dynamite causes it to explode and here we clearly have two events: a lighting and an exploding. It is also obvious that the former caused the latter. But should the features of this example dictate our general theory of causation? Must all cases of causation fit the model of one event bringing about another?

There are two different conceptions of events, one of which allows them to be static and unchanging. According to Kim's (1976) and Lewis' (1986c) accounts, an event is a particular bearing a property at a time, and this doesn't automatically require a change to be happening. But this account seems to fit the notion of fact more than event because we tend to think of it as essential to something being an event that it involves a change or occurrence (Lombard 1986). How this dynamic conception of events is then put to use in accounts of causation is in the idea that a cause is a change that is followed by—and many would say that it produces—another change. In the case of solubility, then, nothing happens if the sugar cube is sat on the saucer, by the side of the tea. It is only when a particular change occurs—the sugar is put into the tea—that the sugar is caused to dissolve. The case of causation is thus understood in terms of a succession of events: the placing of the sugar in the tea and the dissolving of the sugar. There are different theories as to what it is in virtue of which the first event caused the second, as we have seen, such as constant conjunction (Chapter 5) or counterfactual dependence (Chapter 15).

Empiricists have traditionally held an event ontology. What there is, according to this view, is a succession of events in terms of which everything else is to be explained or even constructed. Hence, in Lewis' position of Humean Supervenience, the world is a vast mosaic on unconnected events: just one little thing and then another, and then everything else depends on that (Lewis 1986b: ix). The laws of nature, for example, would be understood as a systematization of that history of events, or, rather, what would be the axioms of a systematization from which the world's history would follow. Probability, properties, and, of course, causal connections would all be seen as dependent on that pattern of events. Empiricists are prepared to tolerate a degree of artificiality in defending their event ontology. Things of other categories would be depicted as logical constructions out of events, such as when C. D. Broad said that a thing was just a boring event (see van Cleve 2001: 64). Not much happens with it.

In causation, too, we maintain that there is a degree of artificiality imposed on the presentation of cases in order to make them fit with the idea that the world is a succession of events. Every case of causation is represented as one event followed by another. If we look at the dissolving of sugar again, an empiricist will say that the cause is the placing of sugar in liquid and the effect is its dissolving. This seems natural insofar as we accept, perhaps tacitly, an ontology in which everything is an event. The idea may gain further philosophical credibility from Aristotle's idea that in every cause there is an 'efficient' aspect: a change that prompts the cause to do its work.

However, Aristotle had a much richer conception of causation than that held by empiricists and while he allowed that there were efficient causes, he didn't think that this is all there is to causation. If we think of why the sugar dissolves—what makes it dissolve—the event of it being placed in water is a pretty uninformative answer. As John Heil has said,[2] someone putting the sugar into the teacup is just the story of how it got there. And that cannot really be the causal explanation of its dissolving because many other objects placed in there would not dissolve. A causal statement of this form— a succession of events—omits some of the vital causally relevant information. It is something about the sugar and the water that explains why one dissolves the other. The cause of the dissolving is not, therefore, it being placed in the tea. It is the mechanism that we saw in Chapter 13, where the ionic bonds are ripped apart by the H_2O.

Anti-empiricists, then, are likely to avoid event ontologies especially when it comes to the matter of causation. One alternative is process ontology, which sits well with the tendential approach to causation that we presented in Part III. But another perhaps more common alternative is a substance ontology. In the latter case, it is substances that are causes, perhaps in virtue of the properties they bear. Whichever ontology one prefers, it is still dispositional properties that are doing the causal work. It seems perfectly natural to say that the water in the tea dissolved the sugar, where the relata of the causal relation are identifiable as substances rather than events. This is more natural than understanding the case as a succession of events and, we have just seen, a more explanatorily powerful understanding too. A substance ontology allows that there are events, it is just that they are not the fundamental things out of which everything else is constructed. But there certainly are events, which can occur when substances undergo change, for example. A process ontology, which we prefer, has an even easier time accommodating events because this is an ontology of constant change (for current thinking on process ontology, see Nicholson and Dupré 2018).

Another difference between those who accept an event ontology and those who hold a different view concerns the temporal nature of causation. Empiricists depict causation as involving a succession of events and the way to distinguish the cause from the effect is that the cause came first. Temporal priority is among Hume's three conditions of causation. But both process and substance ontologies are consistent with simultaneity of causation. This is the idea that for a cause to act on an effect, it must exist at the same time as it. In contrast, in the empiricist view, the first event, the cause, is completed before the second event occurs, hence the cause and effect never exist at the same time.

Critics of empiricism then point out that some of the real causal work can be understood as underlying the manifest events that we are easily able to see and record (Harré and Madden 1975, Cartwright 1989). But the events offer us only a superficial insight into the workings of reality. Bhaskar (1975: 221) calls the underlying work-ings the transfactual generative mechanisms. Our example of the stable orbit of the Earth around the Sun would be such a case, or a spring holding two magnets apart, or

---

[2] In conversation.

a barge sailing smoothly down the middle of a canal pulled by a horse on either side. The forces exerted by each horse are real but neither of them are individually manifest in an event. One horse exerts a force south west, the other exerts a force north west, but the barge's movement is due west. The worldly happening is the factual occurrence, but to understand how it is brought about, one has to understand the generative mechanisms, which are transfactual in the sense that they underlie and produce the facts without displaying their individual natures.

A type of generative mechanism would be one whose manifestation was no change at all. A force north that is matched by an equal force south will result in no movement. The lack of an event is caused, according to our account, even though nothing happens. There can be other cases that look exactly the same but are not instances of causation. When no forces act at all, there is also no movement. The two cases cannot be distinguished in terms of what happens, then, at the level of events; they can only be distinguished according to the generative mechanisms that produce the occurrences or non-occurrences.

16.4 Causation of, and by, Absence

Equilibrium stability cases and changeless effects can be thought of as causation *of* absence. These seems explicable and, in most cases, scientifically knowable. We understand perfectly well why the orbit of the Earth around the Sun is stable. What is caused is the absence of an event: the absence of a change. But the effect is not a nothing at all, even if it is a non-event. There are still the astronomical bodies, real things, between which there is this stable relationship. The effect, then, is a something. What is primarily undermined by this kind of case is the theory in which all effects must be events.

Some see a bigger mystery in causation *by* absence, which is supposedly where 'nothing' is a cause. It is easy to understand the alleged examples. Absence of water kills plants and humans, a missing rail could cause a train crash, an absent gene could cause a birth defect, a lack of confidence could cause a stock market crash (in support of causation by absence, see Schaffer 2004).

The mystery of causation by absence is that an absence sounds like nothing and nothingness ought to be lacking in causal power. It is, after all, supposedly non-existent, so how can it do anything? Indeed, it seems reasonable to say that anything that can act, must be a real something. Being causally powerful might be taken as the very mark of existence. Does that mean we should be realists about absences?

As before, however, the absence of an event is not enough to disqualify something as a cause. There may still be something present, such as an unchanging substance, which is doing the causing, so this is not an absence of everything. A vase could be holding water, for instance. The vase is able to do so without changing. Indeed, if the vase were to undergo change, it might be no longer able to hold the water. The vase causes the water to have a particular shape, the same as the inside of the vase, and to remain in a certain place. In this case, both the cause and the effect involve changelessness: the absence of events. But it is clear that the vase is acting on the water even though it is doing so changelessly. And it is clear that the vase is causing the shape of the water rather than the water causing the shape of the vase. This seems to create no metaphysical mystery, despite the absence of events.

What would be a mystery, however, would be a case of causation where there were no substances, no events, no processes, no properties, and not anything. Examples of this kind will be controversial. Here is one, though, supposedly. A bodyguard is assigned to protect the president. At a crucial moment, he leaves the president's side to visit the washroom at which point the president is fatally shot, the bullet travelling through the exact location where the bodyguard had been standing. It might be concluded that the president's death was caused by the absence of the bodyguard.

However, this can be disputed. Had the bodyguard been in the way, it is indeed true that he could have stopped the bullet. But this is not decisive. Had anyone else been in the way, they too could have stopped the bullet, yet we don't say that all those other people caused the death. Why say, then, that the bodyguard did? One might argue that it was the bodyguard's duty to stop the bullet, rather than someone else's, but now we have cited a pragmatic reason concerning what ought to have happened, rather than what caused the death. Surely the cause of death was the bullet. The bodyguard's absence just explains why it was able to reach the president, but it doesn't cause it.

Dowe (2001) uses a strategy like this to explain why there is no such thing as causation by genuine absence. According to him, when we say that the absence of A caused B, such as an absence of air caused a death by suffocation, then what we have in mind is a counterfactual. In this case, the relevant counterfactual is that if this person had air, she would have lived. And this counterfactual, in the circumstances, sounds correct (unless her death was being overdetermined) given that we have the power to breathe air. So when we say not-A caused B, what we are really thinking is that if A, then not-B. Such a strategy can be adopted for most alleged cases, it seems. A car airbag inflates, for instance, only if a magnetic ball bearing is dislodged from the opening to a metal air-intake tube. The absence of the ball bearing causes the bag to inflate, one might say. Again, however, what is really meant here is simply that the airbag would not have inflated had the ball bearing remained in place. At best, this cites a necessary condition of the airbag's inflation and, as we have seen, necessary conditions are not the same as causes (Chapter 15).

Nevertheless, it might be conceivable that there are real cases of causation of and by absence, though the examples are extreme. The former is easier to accept. If the universe undergoes a huge, cataclysmic implosion, it may be that the effect is that everything ceases to exist. There is at least one theory that this is how it could all end (for instance, Kaloper and Padilla 2015), and if it were to be that way it would count as a caused effect that really is nothing whatsoever. The opposite, causation by nothing whatsoever, would be a literal case of *creation ex nihilo*, the occurrence of which would violate Parmenides' ancient principle that nothing can come from nothing. Nevertheless, there are theories that the Big Bang was exactly such an occurrence.

It is possible, of course, to reject this entire way of setting up the problem. One way of escaping such an anti-Parmenidean conclusion, however, would be to say that the Big Bang was not caused. There is something, when previously there had been nothing, but it was not that the something *came* from nothing, in the sense of being caused by it. A cause must be a something, one could maintain, so the start of the universe must have been technically uncaused. Such a move would allow one to conclude as follows, in summation. There is no causation by absence, where that means an absence of anything at all. There is both causation by, and of, absence of events because causation need not always be of an event or by an event. In some significant cases, the effect is a non-event, such as a stable, unchanging state. At the very least, we have shown that the role of absences in causation is a complicated matter.

17

It All Started with a Big Bang

17.1 The Transitivity of Chain Reactions

Causation does not happen instantaneously. To produce an effect takes time and sometimes lots of time. How much time exactly depends on the type of causal process and the causal elements involved. In history, causal explanations can go back hundreds of years and involve a number of political, cultural, and economic processes leading up to the event in question. World War II, for instance, must be understood as an effect, not only of the rise of Nazism, but also of Germany's situation after World War I, which again had its own causal story. Other causal processes are very fast, such as biochemical ones. Eating sugar quickly results in a raised insulin level, for instance, and when light hits the pupil, the pupil almost immediately retracts. Evolution is a process that could take any time from days, if we are dealing with bacteria, to millions of years for other species.

When making causal claims that span over years, generations, or even millennia, it seems clear that we are not talking about a single instance of causation, but many. How can we plausibly pick out something as the sole cause, if so much else happens over millions of years, before the effect is produced? If we consider the example of evolution, this is an extremely complex process involving all sorts of elements from the microbiological level to ecology and meteorology. The link from cause to effect is in no sense direct here, but happens only via a number of other links and processes.

In this chapter we look at how causes link to their more distant effects. If we backtrack from the effect and consider everything that was needed to get to this point, we might think that there is some kind of necessity involved. Was the effect inevitable, given the causal history? Or was it a mere accident that could very easily have been otherwise? If the Big Bang could happen again in exactly the same way, would all of the world's history be identical to ours? Or entirely different?

How we choose to answer these questions will depend on our philosophical commitments, not only about causation, but about modal matters such as possibility and necessity. Philosophers certainly disagree on these questions and some of these disagreements can be detected in the tension between different scientific theories. A common philosophical assumption is that causation travels down a chain, such as when a line of dominos topples. Given the right set-up, knocking down the first domino should guarantee that all the dominos are knocked over, from the first to the very last. Is all causation like this?

On this picture, causation is what we call *transitive*. A transitive relation is one where if A holds the relation to B, and B holds the relation to C, then A holds it to C. Transitivity plays a central role in Aristotelian logic and classical first-order logic. Aristotle used it for classifications: if all humans are mammals and all mammals are mortal, then all humans are mortal. The question is whether transitivity is a valid inference also for causation. Can we, from the fact that A causes B and B causes C, always conclude that A causes C?

Assuming that this is how causes link to their effects, it makes perfect sense to say that the knocking over of the first domino caused the last domino to fall, even if one needed all the intermediate domino fallings to get there. Does all causation work in the same way?

17.2 Causal Determinism

If causation is transitive in nature, that would make the Big Bang the first cause of every single event in the history of the universe. The Big Bang theory says that this event started off countless causal chains that are still running and will continue to run their course until the end of time. On this theory, the universe is still expanding, and some scientists argue that it will continue to expand indefinitely. This process is thus the longest known ongoing causal process.

An acceptance of causal determinism would be one reason why one might think causes are transitive. Causal determinism is the philosophical view that every single event that occurs is determined by the previous event. So once we have A, we must have B, which must lead to C, and so on. If we start from a certain set of initial conditions, only one next effect is possible, and therefore also the next one after that, and so on. Given the starting point, things could not have been different. In this sense, causal determinism involves a worldly necessity. Causal determinism could then give us an ontological reason for thinking that a cause necessitates its effect. One might also make the opposite inference; from causal necessity to causal determinism.

On this theory, the Big Bang would indeed be the cause of everything that subsequently followed, up until today. It gave us all the initial conditions that necessitated everything else in history, including political conflicts and wars, technological inventions, human choices, and natural disasters. This has led philosophers to assume that causation is the vehicle of determinism, at least for our physical world (for a criticism of this view, see Suárez and San Pedro 2011). Because A is a cause of B and B is a cause of C, each step of the chain is determined by the previous event. But this only works if we also assume that causation involves necessity; that a cause necessitates its effect. We have already seen some objections to this in Parts II and III, but it is the prevailing view in philosophy and in science. It also has some plausibility.

Usually, when we explain a phenomenon, we will backtrack in this way. To understand why an event, z, happened, we will look at what caused it. If what caused z was y, and what caused y was x, and so on, and if each of these steps was necessary in the circumstances, then it seems perfectly plausible to assume that everything that happened actually had to happen, given what came before. Some would call this conditional necessity (see Marmodoro 2016 for a contemporary view and Anjum and Mumford 2018a: ch. 2 for a historical overview), which is different from simply being necessary. This is not unconditionally necessary, because things could have been different. So the dinosaurs are extinct, and it was perhaps unavoidable in the circumstances. But had things been different; had an asteroid not collided with the Earth 66 million years ago, they might have survived.

Causal determinism involves an assumption of conditional necessity. What makes causal determinism relevant for science is its connection with prediction and inference. There are some forms of determinism that are not useful for these purposes, for instance, where everything is predetermined but in a disorderly way that would offer no clue as to what will come next. But other forms of determinism do offer what is needed, filling a perceived explanatory gap. Many scientific models are therefore deterministic, allowing us to predict how the causal process will develop given a certain set of initial conditions. An advantage of this is that we can plot into the model the initial conditions and see how the process will continue into the future. Of course, if some relevant factors are excluded or mistaken, the prediction fails. But if everything is correct in the model, one can trust the prediction based on it. The prediction is thus an inference one makes from the causal history to its future effect.

While causal determinism is a convenient assumption for scientific purposes, then, it has some problems for how the future is understood as something we can influence. If causes determine their effects in the way we have described here, it is not only the past that was fixed by the Big Bang, but also our future, including all our choices. So although the future is yet unknown to us, it would be known to an omniscient being, such as God or Laplace's Demon. The story about Laplace's Demon is an imagined super-scientist with perfect knowledge. If causal determinism were true, it is claimed, then such a super-scientist would be able to predict everything that would happen in the future.

Given for one instant an intelligence which could comprehend all the forces by which nature is animated and the respective situation of the beings who compose it—an intelligence sufficiently vast to submit this data to analysis—it would embrace in the same formula the movements of the greatest bodies of the universe and those of the lightest atom; for it, nothing would be uncertain and the future, as the past, would be present in its eyes. The human mind offers, in the perfection which it has been able to give to astronomy, a feeble idea of this intelligence. Its discoveries in mechanics and geometry, added to that of universal gravity, have enabled it to comprehend in the same analytic expressions the past and future states of the system of the world. Applying the same method to some other objects of its knowledge, it has succeeded in referring to general laws observed phenomena and in foreseeing those which given circumstances ought to produce.   (Laplace 1814: 4)

Popper (1982: xx) refers to this as 'scientific' determinism, because of its central place within classical physics. It is the view that 'the structure of the world is such that *any event can be rationally predicted, with any desired degree of precision, if we are given a sufficiently precise description of past events, together with all the laws of nature*' (Popper 1982: 1–2).

We will now consider whether causation has what it takes to justify this form of determinism.

17.3 Indeterminism: the Default Position?

Popper (1982: 42) denies scientific determinism, which he takes to be an 'utterly unbelievable' thesis. The reason for this is that determinism is such a strong theory that if there were a single event in the whole history of the universe that was not determined, the thesis would be false. Indeterminism should therefore be, on his view, the default view and the burden of proof should lie with the determinist.

To settle whether someone has determinist intuitions or not is fairly simple. Just consider their answer to the following question: if the Big Bang happened again, with the same initial conditions; would the history of the universe be the same? Would everything that has happened in our history happen again? A determinist would say yes, an indeterminist would say no. The question is what makes them go for one rather than the other, which might be a philosophical decision.

We saw that if determinism comes with causation, then all events are necessary given what has happened before. The only way to get out of the causal chain is then to deny causation as a universal principle. One might for instance think that there are some events that are not caused by previous events. Some argue that the radioactive decay of uranium particles works like this: that the decay is spontaneous and genuinely chancy. If this is possible, then the initial conditions after the Big Bang could not possibly settle the future, since there are events that happen accidentally, without prior causes. This is one plausible way to go for an indeterminist who still holds the standard view that causation involves necessity. Indeterminism is then an empirical matter, up to science to resolve.

What if one goes for a Humean world-view instead, that all the world's events are unconnected and accidental: 'Any thing may produce any thing. Creation, annihilation, motion, reason, volition; all these may arise from one another, or from any other object we can imagine' (Hume 1739: 173). Causation is then nothing but the standing of events in a pattern of regularities. Could there be determinism in this type of world, without any worldly necessities? We think so, but it would not be a determinism that comes from causation or one that is useful for the scientist. Instead, it could be that each individual event was determined to happen, including the decaying of a uranium particle and national lotteries, but without any reliable pattern. A neo-Humean could thus be a determinist or an indeterminist, but science would not be able to decide which of these are true. We can call this metaphysical determinism (or indeterminism), since it cannot be settled empirically, even in principle, whether it is true.

A third way to be an indeterminist, is to offer an in-between notion of causation that neither makes events accidental nor necessary. The form of dispositionalism that we propose is one such view and suggests that some things are more likely to happen than others. From this perspective, the initial conditions give us some potentialities. But rather than giving us only one possible next step, dispositions give us many. Even if certain dispositions come together and interact, they will not guarantee an effect but only tend toward it, as discussed in Chapters 9 and 10.

For a causal dispositionalist, life on Earth would be a coincidence in one sense, but in another, the potentialities were already there in the beginning with the causal powers of the elements. Still, some elements, such as gold, could perhaps have been lacking without affecting whether there could be life on Earth, although a number of other things that happened later would not happen without the existence of gold. What this means is that not everything that actually happened before need be necessary for what happened after. Also, it means that things could have turned out differently at every step, even with the same initial conditions. For each stage, some things would tend to be while others would not. Some outcomes might be possible, yet not naturally disposed to happen, while others might seem almost unavoidable. This in-between modality of tendency for causation also allows that many species are possible results of the evolutionary processes, even if only some were actually manifested. Moreover, it also puts some restrictions on what is physically possible, which is only that towards which there are causal powers. For example, not anything is a possible species.

We have argued elsewhere (Mumford and Anjum 2015) that dispositionalism about causation makes it possible to have an open future and free choices, where our decisions can influence what will happen next (see also Steward 2012). On causal determinism, the initial conditions determined which choices we will make and also the outcomes of them long before there were humans on Earth. Besides from rejecting its implicit assumption of causal necessitation, we take causal determinism to be philosophically, intuitively, and empirically unattractive.

17.4 Causation Is Non-Transitive

It seems that causal determinism is the only kind of determinism that scientists need worry about. This is the type of determinism that allows us to make predictions about the future based on knowledge about the present. It also plays a role in causal explanations, showing that what came before was necessary to produce an outcome. Some research programmes seem motivated by a commitment to causal determinism, especially reductionist programmes where it is typically assumed that causal processes at the lower level determine what goes on at higher levels (Chapter 14). Neuropsychology, evolutionary psychology, sociobiology, and the notion of a selfish gene are all ideas that we cannot help but act in the way that we do because of biochemical or neurological mechanisms that are subject to necessary laws of nature. Any illusion of a free action or choice should disappear once we recognize the underlying causal processes leading up to it.

We have seen that causal determinism rests on two assumptions at least. One is that causes necessitate their effects, thus that causation is the vehicle of determinism. The other is transitivity; that causes travel down a chain from the first cause all the way to the final effect. On this view, A causes B, which causes C, and so on, to Z. Ultimately, A is the cause of Z and of all intermediate stages.

But why think that all causation is transitive in this way? There are at least some cases of *prima facie* non-transitive causation. Here is an example. A fire breaks out and produces lots of smoke. The smoke sets off the fire alarm, which then triggers the sprinkler system. Eventually the water from the sprinkler system puts out the fire. Did the fire cause the fire to stop? That would seem a bit odd. Normally, we would think that fire produces more fire, not less. A more scientific example concerns coagulation of the blood. Normally, when the internal wall of the blood vessel (the vascular endothelium) is intact, platelets flow along with the other components of the blood. However, if an injury damages the blood vessels, the platelets get in contact with the internal vascular endothelium, which releases thrombogenic factors. These are proteins that promote the formation of the blood clot (and there are bleeding disorders when people lack such factors). The clot then acts like a cork in the vessel's injury. So, to put this simply: when the skin is cut, it causes bleeding, but the bleeding in turn causes clotting, which then stops the bleeding. But the cut did not cause the bleeding to stop. Rather, it caused the bleeding to start.

What we might say, however, is that the effect wouldn't have happened without the previous links in the causal chain. Going back to the case of the fire, the fire wouldn't have stopped without the water, and the water wouldn't have started if the sprinklers hadn't been turned on, and this wouldn't have happened if the fire alarm hadn't gone off, and the fire alarm wouldn't have gone off unless it was triggered by the fire. This type of counterfactual reasoning shows that there is indeed a causal history of necessary conditions at each step of the causal process. But can we say that, because Z counterfactually depends on Y, and Y on X, and so on until A, then Z counterfactually depends on A?

One way to avoid transitivity for all cases of counterfactual dependence is to distinguish between causes and necessary—*sine qua non*—conditions. It might be perfectly reasonable to say that without the Big Bang, none of this would have happened. But a completely different claim is to say that the Big Bang caused every single event throughout history and will continue to do so in the future. There could of course be some events that are direct effects of the Big Bang. We mentioned earlier the case of the universe expanding. Other events, such as World War I or the recession, do not seem to be effects of the Big Bang, but of many other and much more recent events. So although nothing would have happened if it weren't for the

Big Bang, it still doesn't make sense to say that the Big Bang was the cause of Nazism, the Enlightenment, or the Great Depression (see Chapter 15).

Some philosophers (e.g. Moore 2009: 121–3) try to avoid this problem by arguing that the further back in time the necessary conditions go, the less causally relevant they become. Moore calls this the petering out of causes through time. The Big Bang is then too far back in history to be the cause of anything that happens today. But is this really the case? Sometimes, such as in geology, the relevant causal explanation goes back millions of years, and in history, decades and centuries.

A better reason for not calling the Big Bang a cause of the war can be found by replacing counterfactual dependence with dispositions and tendencies as a way to pick out a cause. An X counts as a cause of Y only insofar as it, in one way or another, tends toward Y. And if a causal factor counteracts an outcome, it is as causally relevant to it as if it contributed to it. But typical of *sine qua non* conditions is that they do neither. The Big Bang is not a cause of any war because it didn't tend towards or away from war. And even if access to oxygen is a necessary condition for insulting someone, the oxygen did not cause the insult any more than it counteracted it.

We have argued that causes don't travel down a chain as a rule. Perhaps many do, but not all. This is why we say that causation is non-transitive, rather than intransitive, which would suggest that causation never goes down the chain. What we do deny, however, is that causes necessitate their effects in a way that can ground causal determinism. With Popper, we take indeterminism to be the only plausible default position. But instead of leaving it to science to provide a case of an uncaused or indeterministic event, we take it that no event is ever necessitated by previous events or causes. If we want determinism, causation will not secure it.

18

Does Science Need Laws
of Nature?

18.1 A Law-Governed Universe?

It may be thought that a conspicuous aspect of science has thus far been entirely neglected. Science isn't just about causes, it could be said; rather, science is fundamentally about the discovery of the laws of nature. We know the gravitation law, for instance, which tells us that the force, F, of attraction between two bodies of masses m_1 and m_2, will be $F = Gm_1m_2/d^2$, where d is the distance between m_1 and m_2 and G is the gravitational constant. It is arguable that the discovery of such a law, which is absolutely general and the same all over the universe, is of far greater scientific significance than knowing particular causal facts, such as that one particle was diverted by another. This is especially so since the generality of the law is assumed to apply not just to actual cases but also to possible cases. The gravitation law, for example, applies to any possible combination of masses and distances and not just those that are instantiated in reality.

The point should be conceded. The generality of the law means that it is useful and applicable in other contexts, hence it permits counterfactual reasoning and predictions. Knowledge of the gravitation law allowed us to put men on the moon but it is not just about the actual behaviour of known objects; it applies to all the behaviour, actual and merely possible, of all objects, whether known or unknown. Thinking of the world in terms of laws allows us to systematize our knowledge. In this we are fortunate. The world seems to accord with a set of principles without which it would be inconsistent and unpredictable. It is natural that within each science we seek these principles. Indeed, it might be thought the very mark of a science that it is able to present its truths in this form. One might question whether psychology, for example, is scientific, if it is unable to give us any strict laws concerning human behaviour. Perhaps the success of any science is a function of the extent to which it is able to do that. The move towards more neuroscience in psychology could be precisely because this is a domain that is more suitable for the delivery of general laws of nature. We find laws in physics and chemistry, but they are less easy to find in biology although genetics and biochemistry might offer some chance. Social sciences and psychology are less able to offer strict laws—anthropology, for instance—and this again might be used to question the status of these disciplines as sciences.

Now it is one thing to note that sciences make appeal to laws of nature. But there is a further and much bigger question concerning whether these are laws that govern the workings of nature. The notion is sometimes deployed of a law-governed universe (Trefil 2002: xxi) but can we take that literally? Is it just a metaphor and, if so, is it an appropriate metaphor? This is a question that also has a direct bearing on the issue of causation because one could also question whether causes are themselves governed by laws of nature. Why, one could ask, do various sugar cubes, in different places and different circumstances, all dissolve in water? An answer might be that they do so because there is a general law of nature that makes them dissolve. The idea that causes are law-governed could thus explain their generality. After all, if one accepts causal singularism, it is consistent with the idea that although causes are real, anything could cause anything, in which case we would be completely unable to predict what will happen. Our world is ordered and predictable and the nomic conception of nature tells us that this is because causal relations, like everything else, are governed by natural laws. If laws don't do this job, it seems that something else must.

We will go on to ask questions about the being, the reality, the substance of laws of nature, and how they are supposed to govern their instances. Before those critical questions, though, the appeal of understanding the universe in this way has to be fully acknowledged. In particular, the attraction of understanding scientific knowledge in lawlike terms should be recognized. A plausible explanation for this attraction is that laws allow satisfaction of a number of the norms of science, especially the norms of systematicity and generality, scientific ideals that were promoted also by Plato and Aristotle.

We read it from the established practices within science that systematicity of theories is one of its norms. If we can mathematize the phenomena and subsume them under laws—as few as possible—then that counts in favour of the science in question. Mathematization might work best with a degree of idealization: for example, where we have a smooth curve drawn through the data points, ignoring any outliers, and expressed as a function. It will be desirable too if the theory is able to unite phenomena that were previously regarded as disparate, for instance, in the way that chemical theory was able to offer a unifying theory of all the many different sorts of reactions that can occur when different types of substance are mixed.

Lewis (1973a: 73, 74), following Mill (1843: III, iv, 1, 207) and Ramsey (1928: 143 and 1929: 150), offers another way in which this systematicity can be understood. We will have a preference for laws that explain as much as possible of what happens in the world. But we also have a preference for economy: in other words, we want as few laws as possible, because this is taken as indicative of their systematizing and unifying power. We would not want, however, so few laws that there are many phenomena those laws fail to explain. There is a trade-off between these two requirements, which we can call strength and simplicity. We will usually seek an ideal balance between the two, then. We will look to explain the phenomena with as few laws as possible (simplicity) that are able to explain enough of the phenomena (strength). So, for example, if we could explain 98 per cent of the world's facts with just four fundamental laws of nature, we might not think it worth adopting a fifth fundamental law if it explains only a further half a per cent. Doing so would complicate the theory for relatively little gain. We might indeed prefer to seek a different set of four fundamental laws that explained everything. As always with the norms of science, however, it may still be debatable as to how we should satisfy them.

The second norm of science relevant to lawlike conceptions of science is generality. Science aims to provide knowledge of the world that is not merely local, concerning particular matters of fact, but which is universal in that it concerns all the particulars of a certain kind. The 'success' of Boyle's law, that the pressure on a gas increases as its volume decreases, requires that it applies to all gases. If it applied only to hydrogen, it would be of less interest and use to us. If it applied only to samples of hydrogen in a particular place and time, under rare circumstances, it would be even less use to us, or none at all. As a rule of thumb, the more general the law, the better, which explains why Newtonian mechanics has been so successful for so long. With just three laws of motion and a law of gravitational attraction, the movements of almost every body in the universe can be explained. Einstein showed that there were some cases to which Newtonian mechanics didn't apply and that can be taken as a reason for preferring Einsteinian physics to Newtonian.

Now it might be noted in the case of Boyle's law that although it has generality, it is also stated in a way that restricts its application. It applies only to ideal gases and these exist only theoretically. In an ideal gas, the particles have no interactions other than collisions, so there are no intermolecular forces at work, for instance. How there can be an advantage to having a law that is both general but restricted in this kind of way is an issue that we will have to address. It is clear why generality is useful, though. If we have a law about all electrons, then we can make predictions even about the electrons we have not yet observed. We can also use general laws for explanation. If, for example, we know a law that any rat eating more than eight pellets of poison will die, then we can explain why a particular rat died that had eaten eighteen pellets of poison (from Popper 1972: 350, 351). Again, though, we will also need to address the issue of how laws of nature are supposed to explain the phenomena within their scope.

18.2 The Being of Laws

There are some basic questions of metaphysics that can be brought to bear on the notion of law of nature and the law-governed universe (a detailed study is in Mumford 2004). We will see that these impinge very directly on the issue of causation; indeed, some use the term 'causal laws', thereby uniting the two notions. The two big questions that we address here are: what is the nature and existence of laws and how, if at all, would they explain their instances? The answer to the first of these questions is likely to determine the answer to the second.

There are two main theories of what a law of nature consists in. Neither of these is adequate, however, which is why we will offer a third option as well.

The first answer is the Humean one. In such an account, a law is nothing more than a statement that summarizes or systematizes what actually happens in the world. We might observe the temperatures at which water boils, for instance, and note a similarity among the instances. We can see some variation, according to altitude, but if we plot the data then there is a clear-enough pattern. Water boils usually around 212°F, and on the Celsius scale we assign the mark of 100°C to this significant point. A norm of science is, of course, to report the facts and this can be applied to the construction of our theories of laws. We gather all the available facts, look for a pattern, and then sum it up neatly in a law statement, perhaps with the aid of some simplification and idealization (see Chapter 6). In simple terms, this gives us the essence of the regularity view of law (Psillos 2002: ch. 5).

Something to note immediately, however, is that this does not give the law of nature any real existence as an entity in the world. What exists, what is real, is the regularity: the set of facts that exhibit a pattern. The law is just a statement of that pattern. The law is not an ontological existence in its own right, though the pattern is. It follows that laws, conceived in this way, cannot govern their instances. It cannot be, for example, the law that makes water boil at 212°F, if that law is just a summary of the fact that it usually does. Rather than this being an embarrassment for Humean theories, however, empiricist philosophers will usually see this as a virtue. They want a non-governing conception of laws (see Beebee 2000) because they deny that there is any necessity in nature.

One might be prepared to accept this Humean position, but we do need to understand what comes with it. The account means that laws cannot, after all, explain their instances. We said that we could explain the death of a rat with reference to the law that any rat eating more than eight poison pellets dies. But if the law is just a summary or systematization of the fact that rats eating that many poison pellets have died, then it doesn't explain those deaths. It is merely a statement that they *did* die, which is not to say *why* they died. In other words, the existence of the instances explains why there is a law, according to this view. It is not that the law explains the instances. Humeans may be happy to embrace this conclusion and they can do so consistently. But it does mean that laws don't add much to our understanding except, perhaps, that we can see these rat deaths as part of a pattern instead of being isolated incidents. Humeans will be keen to point out, however, that the existence of a pattern does not mean that there are any real necessary connections underlying it.

Related to this, and perhaps more damaging, is the consequence that laws would lose also their predictive role. The view tells us that the law is just a summary of what has happened. But given that there is no necessity in anything that happens, there is no reason why unobserved cases should be like those observed. We arrive at the problem of induction. The problem means that we have no rational basis on which to make predictions because the facts of what has happened do not, within this theory, support any conditional reasoning. Hence, on the basis of the law, one might think it rational to suppose that if a rat were to eat nine poison pellets in succession, it would die. Hume himself can say nothing more than that it is a custom or psychological habit that we draw such inductive inferences but it is strictly speaking not rational to do so.

Because of these perceived inadequacies of the Humean view, some have sought more robust conceptions of laws in which they can be genuinely governing. Governing conceptions had been critiqued by empiricists because of the associations with the notion of a natural lawmaker, which meant that the main governing conception was a theological one (see Ruby 1986). Just as the laws of the land are made by government to regulate the behaviour of its citizens, so one might think that a divine lawmaker has made laws that govern the behaviour of natural objects. When allowing himself his more philosophical moments, Newton seemed to think something like this (Newton 1687: Preface). One might then wonder whether any sensible, non-theological account can be given of governing laws.

A modern, non-theological attempt came from Armstrong (1978, 1983). Armstrong recognized the above failings of the Humean conception of laws and wanted a theory in which the law could be said to determine its instances and at the same time support counterfactual inferences. He did so in the following way. First, unlike the regularity conception of laws, he took it that law statements were not simply generalizations—universal quantifications—over the instances. Laws should not be understood as having the form *all things that are F are G*; for example, everything that is an iron bar and heated expands. As we have seen, that is only about the actual iron bars that have been heated. It says that they have all expanded. But it tells us nothing about those that have not been heated.

However, if the law directly connects the properties or universals involved, F and G, rather than the actual particulars that bear those properties, *a* and *b*, then the law will apply to anything that were then to instantiate F and G. In other words, instead of the law applying to various particular things that happen to be iron bars, it applies to anything that has the property of being heated iron. Furthermore, the law consists in the holding of a relation of natural necessitation, N, between that property and the property of expanding. Thus, the true form of a law statement is N(F,G), which means that being F *naturally necessitates* being G.

For Armstrong, properties such as redness, squareness, circularity, being heated, and being expanded are universals, which means that they are identical in their instances. Everything that is circular is exactly the same in respect of its circularity. And N(F,G) is itself a universal: it's a higher-order universal because it's a relation whose relata are themselves universals. Universals come in two kinds: properties and relations, both capable of having multiple instances. Because N(F,G) is a universal, hence identical in all its instances, this entails that anything that is F must be G. So a law of nature determines its instances, on this account, because everything that instantiates F must also instantiate G. It thereby instantiates the law, not just by instantiating F and G; but specifically because there is a connection between being F and being G. And what would the instantiation of that connection be at the token level? It would be causation, of course.

This solution is neat and ingenious. Laws, as universals, exist in their instances, those instances being particular tokens of causation, such as the heating of this iron bar causing it to expand. And we need an account where there is something more than Humeanism so there needs to be more than simply the occurrence of two events in order to say that one causes the other. Armstrong's answer is that one thing causes another if and only if it thereby instantiates a law of nature. Not all conjunctions of events do. Indeed, even a constant conjunction of events need not, where it is accidental. All spheres of gold are less than 10m in diameter but not because any law makes it so. An answer is thus provided for how accidental regularities differ from genuinely lawlike ones.

18.3 The Powerful Alternative

Armstrong's view has faced much criticism. It relies heavily on his immanent realist metaphysics of universals, which not everyone accepts. And the relation of natural necessitation has been thought of as a *deus ex machina*. Just calling it natural necessitation doesn't ensure that it is (Lewis 1983: 40). We will set aside those criticisms, however, so that we can concentrate on two other issues. First, we question the wisdom of understanding the world in terms of necessitating laws. Second, we will question whether the causal powers of particular things really do stand in need of governance from above. We will then be able to offer an account in terms of causal powers in which laws of nature are redundant. This is not because powers can fill the same role that laws were intended to have. If that were so, you might think of the powers and laws theories as equivalent. It is more that with the right account of causal powers, you can see that there is nothing else needed to explain the order and tendential regularity of the world.

We have explained already why we object to a necessitarian view of the world. To satisfy the norm of generality, laws are depicted as strict and inviolable, within governing conceptions, and they thus have a problem allowing exceptions. Although Armstrong's law is not of the form *All things that are F are G*, it is nevertheless supposed to entail it. And yet, we have already argued, there are no such *de facto* exceptionless regularities to be found in nature, at least not where causation is involved. Boyle's gas law seems to hold universally but only because it applies to some idealized kind of gas when, in reality, the truth is more messy. Likewise, the gravitation law is true in a sense, of course, but it is also an idealization in that any attractive force between two bodies must be abstracted from the many forces acting upon those two objects, given that there are always multiple bodies, other forces, and motions present. As we have already argued (Chapter 10), therefore, we prefer an account of *what tends to be* in nature, which we think fits the world's empirical messiness better (see also Cartwright 1999).

But how, then, could one account for the explanatory utility of laws? Even Boyle's law is useful, despite its content nominally being about ideal gases. The best approach is to accept the explanatory power of laws and offer an account of their usefulness. Accepting laws as mere statements that summarize or systematize nature seems correct, and to that extent we agree with the Humeans. But we disagree with Humeans over what they are summaries of. Humeans think of the laws as summaries or systematizations of the actual events. They favour an ontology of events, as we saw in Chapter 16. There were rational grounds, however, for accepting a richer ontology of underlying mechanisms that generated those events, and we can think of these as causal powers. Furthermore, it makes more sense to think of the laws as statements of what these powerful mechanisms are, rather than statements of actual events, because this would be a better way of explaining why laws about ideal objects and circumstances are nevertheless informative.

The ideal gas law, for instance, is useful because it tells us something about the natural inclination of gases, for their pressure to increase as volume decreases, which is still there in real gases even though there are other forces working upon them that can render the law factually inaccurate in some cases. Similarly, the force of attraction between two bodies will rarely be factually as stated by the gravitation law. Other bodies will be exerting attractions on those objects and there can be other attractions and repulsions just between those two bodies: electrostatic repulsion, for instance. But the law is correct if taken as a description of the tendency or disposition of the two objects. Such an interpretation is entirely natural because, after all, force, attraction, and repulsion are all clearly disposition terms. That there is an attraction tells us nothing about what actually happens but it does tell us something about what is disposed to happen, so this makes the dispositional reading of this law the obvious one.

It might be worried that a tendential view of laws makes them immune from falsification. At least if a law is strict, one contrary instance falsifies it, as Popper (1959) claimed. But laws that concern tendencies, it might be objected, could not be falsified by a counterinstance, since they do not entail an absolute generality. However, it need not be granted that laws based on tendencies are unfalsifiable. Laws, we suggest, should be understood as attributions of causal powers to things, for example, that masses attract or that a particular substance is poisonous. Such a putative law would be false when the thing in question did not have that power. Knowing that something does not have a power is not a simple matter, we concede. In particular, we would need to make sure we have ruled out a similar-looking possibility that something has a power but it is being counteracted. Nevertheless, it is not impossible that we tell the difference between these two cases. By looking at the totality of evidence, including that gained from intervention and experiment, it might be possible to conclude rationally that there is no power present in a particular case.

Perhaps we can isolate the case and run various tests. But we might have to say more about the overall evaluation of evidence, in its various forms, before this view can be fully persuasive.

Such an account also offers some hope for the vindication of sciences that struggle to offer strict laws of nature. The strictness of many of the laws is illusory in that what is strict is that there is an inclination, comprehensible through isolation and idealization, and this is less acceptable in other sciences. We expect psychology and economics to tell us about real people and economies, not just ideal ones. But even then we might observe that real people tend to blush when embarrassed or that increases in price tend to lower demand.

The powers view also gives us some explanation of why generalities hold. Ours is not a Humean world of contingency, where anything goes. Poison really does have a power to kill rats and will tend to do so. The law is thus a summary of the nature of the poison (and nature of rats, as mutual manifestation partners to that poison). It might yet be wondered, though, why all things of the kind behave in the same way. Why are they all similarly empowered? Some think that laws of nature are what explains this (Everitt 1991). But one can turn this question on its head. There is no mystery in why things of the same kind are similarly empowered: it is *because* they are similarly empowered that they are classed as belonging to the same kind. This is the view of dispositional essentialism (Ellis 2001). Hence, it is no coincidence that all electrons have unit negative charge. Having unit negative charge is one of the features that makes something an electron. It is an essential property of being an electron.

Causal powers are thus capable of furnishing the world with order, adequate predictability, and explanatory coherence. We would maintain that they do so more credibly than governing conceptions of laws. They do so in a different way: in a dispositional rather than necessitarian way. We think this latter point important. Powers are not merely filling a law-shaped hole in nature. There is no such hole to be filled. But we have given an account in which laws of nature, understood as useful statements underwritten by the real powers of things, can have perfect validity. Our account of the world is not law-governed from above; it is driven, rather, by really empowered individuals.

PART VI

Probability

19

Uncertainty, Certainty, and Beyond

19.1 Mathematical and Natural Probabilities

Any discussion of theories in science, including causal theories, will eventually come up against the question of probability. There are various reasons for this, concerning degrees of belief, inconclusive evidence, and the chancy nature of the world itself.

Our belief in a theory could be partial where we think it is probably true but we by no means are certain. A belief in a theory might be tentative, or perhaps strong but still short of certainty. We can also think of evidence as admitting degrees. On the balance of evidence, a theory could be more probably true than not but without any warrant for saying it is definitely true or definitely false. There is a stronger view that probability is not just about our beliefs and knowledge but can also concern the way the world is: that it is inherently chancy. In this part, our aim is to disentangle these separate notions of probability.

It is not our only aim, however. Perhaps more importantly, we want to make a clear distinction between the *mathematized* and *natural* conceptions of probability, which corresponds to Mellor's (2005: 18) distinction between pure and applied probabilities. Our contention is that the mathematical conception of probability, while no doubt being an immensely helpful tool, does not match exactly with the natural state of probabilities: neither as we reason about them nor as they are. Given that our concern is with causation in science, any discussion of probability should be about probabilities in the natural world. Furthermore, although the mismatch between these two conceptions appears slight, one being an idealization of the other, an assumption that the natural world behaves according to a classical conception of probability can give us a misleading image of the workings of causes.

Chapter 19 will focus mainly on the epistemic topics of probability as credence, or degree of belief, and evidential probabilities. Chapter 20 will then focus on probability as a worldly phenomenon, in which we will outline a distinctive account of propensities in opposition to frequentism. In Chapter 21, we then show how this account of natural propensities would require revisions to the orthodox treatment of conditional probability.

19.2 Belief in Theory

The first question, when addressing the topic of probability, is whether we are talking about probabilities as a feature of the world or merely as an expression of our relative states of ignorance, such as doubt and knowledge or the balance of available evidence. If a theory is true, for instance, it is still possible that the available evidence is unable to indicate that it is so conclusively. Someone might put a coin behind their back and then produce two closed hands, asking you to say which one contains the coin. The evidence does not determine whether it is in the left or right hand. Evidentially, then, one could say that there is a 50:50 chance that it's in either. But this does not of course mean that the world itself is chancy. The coin is entirely in one hand or the other. The fact of its location is ontologically definite; but the evidence underdetermines which fact is true. Credence is something different than chance, too, and concerns subjective beliefs about probability. Suppose you have an implicit bias towards always choosing the left over the right. You might have a higher degree of belief that the coin is in the left hand than the right simply because of this personal bias, even though the evidence does not support this probability assessment.

The other possibility is that the world is inherently chancy. A determinist thinks that it isn't (Chapter 17), which effectively means that the real chance of everything is either $=0$ or $=1$; that is, $=0$ if it happens to be false and $=1$ if it is true. But indeterministic accounts of the world gained greater credibility during the twentieth century due to developments in quantum physics that invoke real-world chanciness. Radioactive decay is the well-worn example in which, for instance, a radium atom could decay at any time during its existence and there is nothing that determines the exact moment at which it does. Not every theory offers this interpretation but some do. Nevertheless, the decay is not entirely random but, rather, probabilistically constrained in the sense that there is a 50:50 chance that a radium atom will have decayed after a period of 1600 years. Because of the indeterministic nature of the decay, the time of decay can only be given as a 'half-life', as above, with radioactive atoms of different kinds having shorter or longer half-lives.

This is just one type of real-world indeterminacy and it might be considered exceptional insofar as it concerns atomic entities. Nevertheless, the case could be significant because it involves what look like fundamental phenomena, hence the indeterminacy cannot be reduced away into something deterministic (although a hidden variable theory says it could, see Bohm 1957). The case is also significant if one accepts that causation is bottom-up and that indeterministic phenomena are at the bottom of everything. However, one cannot automatically infer from the bottom level being indeterministic to all higher levels being indeterministic. The theory is consistent with determinism holding at higher levels. Applying the conclusion of Chapter 14, the deterministic nature of higher-level phenomena could be an emergent property of them.

These considerations have a direct bearing on our acceptance of causal theories. Those theories might be posed in probabilistic terms either because they are inherently chancy, or because the evidence is inconclusive, or that we want to express a degree of assent (or doubt) in a theory that is short of certainty. One might also think that such doubt, uncertainty, or judgement of probability accords with the spirit of science; namely, the norm that one should apportion belief to the evidence and no more. One should not believe more than what the evidence supports. Science should not be dogmatic and it should not draw unwarranted inferences. Hence, a claim such as that human activity has caused climate change should be given a probabilistic weighting, even if it is a very high one. After all, can one ever have absolute scientific certainty? Isn't there always at least some degree of doubt, of at least 0.1 per cent, for instance; in which case the degree of belief should never be more than 99.9 per cent?

We find this sort of thinking within Bayesian reasoning, for example, which arguably can apply to all of credence, evidential, and worldly probabilities (Bayes 1763, Laplace 1814, Earman 1992, Pearl 2000, Swinburne (ed.) 2002). At the heart of Bayesianism are some attractive norms, for example, that one should update one's beliefs in the light of new evidence. Hence, if we have a prior belief about the probability of a theory, T, and then encounter new evidence that is relevant to it, we should either increase or decrease our degree of belief in T, depending on whether the evidence supports T or not. This norm is sensible and it accords with the scientific spirit in that without it a theory would seem unanswerable to new empirical data. If empirical evidence under no circumstances could affect our degree of belief in T, then one can only assume that T no longer qualifies as an empirical theory but instead is a postulated truth. Bayesianism offers a normative account of how beliefs should be updated in line with new evidence, where the probability of the theory T, given the empirical evidence E, is equal to the probability of E, given T, multiplied by the prior probability of T, all divided by the prior probability of E. That is, $P(T \mid E) = P(E \mid T) \cdot P(T)/P(E)$, or as the formula states:

$$P(A \mid B) = \frac{P(B \mid A)\ P(A)}{P(B)}.$$

A second norm of science, with which Bayesian reasoning accords, is that real causal connections tend to be revealed in the recurrence of events. Price, posthumous editor and publisher of Bayes' (1763) essay, said almost exactly this in his introduction: 'it shews us, with distinctness and precision, in every case of any particular order or recurrency of events, what reason there is to think that such recurrency or order is derived from stable causes or regulations in nature, and not from the irregularities of chance' (Price 1763: 124). This reading of Bayes would even be consistent with a propensity interpretation of probability. It would allow a process of weighing evidence, updating prior assessments of probability in the light of evidence, and thus tending toward an accurate assessment of the strength of the worldly cause involved. The reasoning would be that causal propensities, which we consider more in Chapter 20, will tend to manifest themselves in stable patterns of events (see for instance Popper 1959 and Gillies 1973). In contrast, if the world were as Hume describes, then existing and new evidence would tell us nothing at all about what is more probable or not. We accept Price's point that the unfolding of events, especially if they exhibit a degree of order, is evidence—some evidence, even if it not complete and certain—of the existence of underlying causes producing that order or recurrence. Jakob Bernoulli had a similar idea (Gower 1997: 90). You could assume a hypothesis that A causes B and then look to the evidence in events to see whether that hypothesis was confirmed or not.

We agree with the view of science, central to Bayesianism, that theories should be accepted only tentatively, with a degree of uncertainty that is sensitive to the acquisition of new evidence. There will be qualifications to follow. But the mathematical model by which any such probabilistic reasoning is to be understood must be treated as a separate matter, for reasons we will now present.

19.3 Modelling Chance

Ask someone what the chance is of a fair die landing on a two when rolled and they will almost certainly say one in six. Nevertheless, the answer is false if it is meant as a statement of real-world natural probability. This is not simply because determinism could be true, in which case one might maintain that the real probability is either one or zero, depending on the exact initial conditions. Let us take the question as a matter of mere epistemic probability, if it helps. Since a definite calculation of the die's final resting position is virtually impossible, a statement of one in six as an epistemic probability seems rational. This is, however, because thinking in terms of classical probabilities, as axiomatized by Kolmogorov (1933), is entirely normal. The reasoning progresses from the assumption that there are six equiprobable outcomes to consider, landing on a two being one of them, and thus the probability is one out of six.

However, the assumption that there are only six possible outcomes is itself artificial. These may be the six most likely outcomes but to say that they are the only possible outcomes tells us that we are considering an idealized model of what will happen. In reality, there are a number of other possibilities. These may be vastly improbable, of course, but taken together they must make the chance of the die landing a two somewhat less than one in six. The die might break apart, for instance, be caught after its first bounce, or be incinerated in an explosion just before it comes to rest. It might also balance on an edge if it leans against something. There are these other possible outcomes because we are now dealing with physical probabilities, not merely an abstract simplification of them.

Classical probabilities were good for understanding games of chance, where the hand could be voided in the event of the dealer dropping the cards, for example. They allow the gambler to calculate a one in four chance of picking a diamond off the top of a full deck and a one in thirteen chance of picking a queen of any suit. The theory can be used as a heuristic, therefore, and is perfectly serviceable for predictions and thence for wagers. Such thinking is useful scientifically, too, especially if one can weigh the possible outcomes. Thus, one might estimate four possible outcomes from an intervention, though one of them being twice as likely as each of the others. One can then calculate a 2/5 probability for this outcome and 1/5 for the others. Perhaps predictions made on such a basis, even when applied to the real world, will tend to be right, such as if one has to predict a distribution over a series of trials.

It was Daniel Bernoulli who suggested that hypothetical reasoning followed a mathematical structure, which allowed probabilistic reasoning based on an idea of degrees of certainty (see Gower 1997: 94). This paved the way for Bayesian reasoning in science. Galileo's idea that the book of nature is written in the language of mathematics (Galileo 1623) supports this view, which could be traced as far back as Pythagoreanism. In contrast, d'Alembert (see Daston 1979) thought that reasoning didn't have a mathematical structure. Which view is right?

19.4 Certainty and More

Daniel Bernoulli's notion of certainty suggests a maximal degree of belief, =1. This is also reflected in the credence interpretation of Bayes' theorem where the maximum you can believe something to be probable is degree =1. But degrees of belief might not follow this structure. For one thing, you might not be able to put numeric values on degrees of belief at all. Plato proposed a division of epistemic states into belief (*doxa*) and knowledge (*episteme*) (Plato, *Republic*: V), for instance, which are distinguished qualitatively rather than quantitatively. Added to this, we could say that we are less or more certain of one thing than another. But as Mellor (2005: 18) explains, these do not necessarily demand assignment of numeric values. We can simply make comparative judgements without assigning quantities to belief. For example, you might be just as certain of the theory of evolution as of the theory of gravity, or more so, or less so. In contrast, it may seem phenomenologically implausible, and indeed arbitrary, to say that one belief has a 0.2 degree while another has a 0.63 degree. How is any such precise degree known or experienced?

Again, though, a reply could be that a framework of classical probability theory applied to degree of belief is relatively harmless, even if it is psychologically implausible. It is simply a useful heuristic, one might say, to assign values within a range where 1 = certainty that something is true and 0 = certainty that something is false.

However, we cannot accept this kind of pragmatic reply in its entirety because we think there will be significant cases of belief that do not fit the model but which are nevertheless relevant to evidential reasoning within science. The cases we present may sound challenging if one's thinking remains within the standard model of degrees of belief. We contend, nevertheless, that they are perfectly intuitive and acceptable if one considers natural phenomena and not just the mathematization of credential probability. Put in its simplest form, we propose that there are some matters about which one can be *more than certain*. This could apply both to being more than certain that something is true and more than certain that something is false. It sounds wrong if one models beliefs mathematically in such a way that certainty occurs only when degree of belief = 1 and there is no belief with degree >1. But this, we maintain, exposes a limitation of the model of belief.

How, then, can a belief be more than certain? We maintain that it can be so in the following way. Certainty is where you believe something completely and without reservation, let us suggest. There can be cases where there is more than enough reason to believe something in that you would still believe it completely and without reservation even if there was less evidence, within limits. Suppose that in front of an audience, and televised, Thomas is seen strangling Richard. Everyone is convinced that Thomas is guilty of murder and he is brought to trial at which a mountain of evidence is presented against him. The jury is already persuaded of a guilty verdict when it is revealed that some fingerprints taken from Richard's neck did not belong to Thomas. Nevertheless, they might still judge it was an absolute certainty that Thomas killed Richard. In other words, there was more than enough evidence to convict Thomas. The judgement of certainty was able to withstand the removal of this one item of evidence. Of course, assuming the jury to be rational, the verdict could not withstand removal of all evidence. But this is nevertheless presented as a plausible case where at least some could be subtracted with certainty remaining intact.

This type of case should sound familiar. It can be likened to the phenomenon of overdetermination, already considered in Chapter 15. There is, thus, some redundancy in the determination of the certainty of belief by the evidence. Now if one assumes a Bernoulli/Bayesian framework for credence, in which 1 is the maximal value of belief, then to follow Bayesian reasoning, one should lower one's posterior probability in light of the evidence that counts against it. If Thomas strangled Richard, one would expect his fingers to have left their mark. The fingerprints belonging to someone else tells us to revise our degree of belief downward, which means it has to then be <1 and thus less than certainty.

It is easy to think of examples that more clearly concern scientific theorizing, though exhibiting the same principle. Some evidence might be produced that counts against the theory that smoking causes cancer. Perhaps someone smokes for sixty years before dying of other causes and no cancer is found in his body. Such evidence counts against the causal theory under consideration and thus one's degree of belief in it should be revised downward. However, what is not clear is that this automatically requires that it be revised downward enough such that it would fall below a threshold for certainty. It would seem perfectly rational to say that there is more than enough accumulated evidence, of all kinds, for us to believe that smoking causes cancer. Even without the confirmation found in this one additional case, existing evidence could be more than enough to place belief beyond the threshold for certainty. Such a belief is held strongly enough that it can survive the subtraction of some of its evidence or the discovery of some contrary evidence.

The problem here is that degree of belief has been modelled on a bounded scale with 1 as its upper limit and 0 as its lower limit. But what is the reason to assume an upper limit to strength of belief, or a lower limit? The lack of a lower limit is tantamount to there being more than enough certainty that something is false, such that one would still be certain it was false even if some evidence in its favour came to light. For example, one might be certain that there are no ghosts of dead people inhabiting our world. One could still believe this even if someone produced a photo that looked like a ghost. This shows that one's lack of belief in ghosts is so strong that one would dismiss apparent evidence in favour of them rather than accepting their existence.

An opponent might say that certainty is certainty, *simpliciter*, so that a notion of more-than-certainty is of no service. But we have already shown why this claim does not hold up. It is important to distinguish between 'mere' certainty and more than certainty because the latter, but not the former, can survive the loss of some supporting evidence. At the very least, therefore, there is a counterfactual difference in the epistemic states of one who is more than certain and one who is merely certain. Because of this, then, it will not do simply to impose an arbitrary threshold for certainty, somewhere <1, of degree of belief; for example, saying that certainty is any degree of belief >0.9. For even this suggests there is still an upper boundary to extent of belief, whereas it seems more plausible to allow instances where one can find more reason to believe something of which one is already certain.

It appears, then, that even if it is possible to assign numerical values to beliefs, those beliefs would still have to be measured on an unbounded scale rather than the bounded scale of classical probability theory. There need be no upper or lower extent of belief. A strongly held belief that is held with certainty could be held even more strongly if there is new supporting evidence. It might be rational to acquire such evidence, even though it is redundant. We now know more about the mechanism that connects smoking with heart disease, yet it was clear that smoking caused heart disease even before this mechanism was found (Gillies 2011). One might wonder why the mechanistic knowledge was required, given that the link was already known. But we are now in a position where we have a variety of overwhelming evidence in favour of the causal hypothesis. Hence, there would still be reason to believe it even if some correlation data from the past, linking smoking and heart disease, were at a point invalidated or superseded.

19.5 Remaining Doubts

It might be questioned where this leaves the scientific spirit, discussed earlier, in which there should always be some room left for doubt. After all, no scientist claims to be omniscient and there have been cases in the past where something seemed more than certain but it turned out to be false, such as the geocentric theory of the universe. The epistemic case of more-than-certain beliefs is, we shall see, a case of overdisposing. We will consider this in detail in Chapter 20, including an explanation that reconciles the dispositional notion of there being more than enough for an effect with the less than 1 classical probability of its occurrence. This is important because, as we have argued, causes never necessitate their effects. It is, we will go on to argue, possible to square these two claims because degree of power and degree of belief are not being measured in the same way as classical probabilities.

This view of belief may be all well and good in theory but is there anything to recommend it within science? Is it an isolated epistemic theory, for instance, or does it relate to other significant matters in philosophy of science? The account fits science, we claim, in particular by making sense of the complex relations between confirmation, falsification, and theory change. Contrary to Popperianism, one does not, and should not, automatically abandon a theory in the light of one falsifying instance. If the theory is otherwise useful and seems to work, then it should be retained. One could bracket the exception as an anomaly to be explained away at some later point (Lakatos 1970). One could even tolerate a number of such anomalies. Although we are not committing exactly to this framework, we can at least show that it coheres with it. If one is working within a theory, a paradigm, or research programme, one can have more than enough supporting evidence to believe it. The evidence can overdispose towards the truth of the theory. And then the theory can be retained in the face of countervailing evidence. However, as shown by Kuhn (1962), it is possible that the counterevidence accumulates so much that it is regarded as too strong for the theory to be maintained. Such a theory moves into a phase of crisis and we are ripe for theory change: a scientific revolution. Thus, a theory can be held and believed for some time in the light of contrary evidence; but without being entirely immune to empirical refutation. We are more than capable of rejecting a theory on empirical grounds but not on the slightest of pretexts. The account of belief that we have offered makes sense of this.

We began with a very general claim, then, that natural possibilities did not have the same structure as the mathematized possibilities of the classical theory. The latter is an idealization of the former which is in many respects useful. However, the way in which it conceptualizes beliefs is inaccurate in some ways and produces inadequacies when it comes to understanding the role of degree of belief in relation to assessments of probability. Having argued this for the case of credence, we will now move on to consider objective probabilities more generally with the case of real-world propensities.

20

What Probabilistic Causation Should Be

20.1 Real-World Probability

Irrespective of the evidence and our beliefs about what is probable, there is the separate question of what the real-world chance is of something happening. To a strict determinist who thinks that everything that occurs does so necessarily, this may well be a non-question. To someone who staunchly believes so, this chapter might be of small value. It is, however, possible to take much of what we will say as if it concerns only epistemic possibility. But we also hope that, in a scientific spirit, the determinist is willing to entertain the hypothesis of at least some worldly indeterminacy. As Popper (1982: 42) has said, indeterminism really ought to be the default position, given that determinism is such a strong thesis. Determinism is true only if every event in the world's history is necessitated and it is hard to find compelling empirical evidence to support that view.

In this chapter, we aim to develop and defend a distinctive propensity interpretation of worldly probability. We will distinguish it from other propensity theories and, of course, rival frequentist theories of probability. We then show how it accounts for chance. Single-case propensities are the grounds for the facts of chance, we argue. It is vital also to separate the notions of propensity and probability, which other propensity theorists bring closer together. The idea of epistemic humility will also be introduced here, as it pertains both to knowledge of propensities and, as explained in later chapters, causation generally. The framework we offer permits talk of genuinely probabilistic causation. This is where a cause, instead of necessitating its effect, probabilifies it and, on some occasions, may produce it. Unlike other causal realists, including propensity theorists, we do not think that these probabilistic causes must resolve into necessitating causes in order to produce their effects. They can remain irreducibly probabilistic. Such a view then validates a notion of probabilistic causation in science.

20.2 Frequency, Chance, and Propensity

There are two main interpretations of chance: frequentism (e.g. Venn 1876, Reichenbach 1949, von Mises 1957) and propensity theory (e.g. Peirce 1910, Popper 1959, 1990, Mellor

1971, Giere 1973, Gillies 2000b, Suárez 2013, 2014). The simplest way to understand the difference between these views is as follows. A frequentist says that the facts of frequency of occurrence determine the facts of probability. The propensity theorist says that there are real-world properties, called propensities, that produce any such frequencies of occurrence. The propensity theorist, then, thinks the exact opposite of a frequentist.

An example will help to illustrate this difference. During the Ebola epidemic in West Africa in 2014, as a statistical fact, 39.5 per cent of those who contracted the virus died from it. One may well then reason that if you visited the region at the time and were to contract the virus, there would be a 39.5 per cent chance of death and 60.5 per cent chance of survival. This conclusion is *prima facie* attractive and, for all we know, may be true. The question, however, is whether this fact of frequency of occurrence—that 39.5 per cent of those with the virus have died—is what determines, dictates, or defines the chance of death as being 39.5 per cent. A frequentist says so. Frequency of occurrence fixes chance. A frequentist could argue that there couldn't be anything else that determined the real-world chance other than the actual occurrence of events. Furthermore, in this case there was a sufficiently large sample from which to draw a conclusion as to real chance. It is not as if one is generalizing from just a handful of cases. There were actually 28,652 people who contracted the virus. What more could the fact of the chance of death consist in other than that 39.5 per cent of those cases were fatal?

A propensity theory, in contrast, maintains that frequency of occurrence *may* be evidence of the real facts of chance but need not be. Frequencies could reveal the presence of a propensity—and be evidence of the strength of that propensity—but they do not define or determine it. The reasoning here is that any such frequency of death among those with Ebola is the outcome of lots of individual people having the virus and going through periods of illness, some of them surviving that illness, and some of them not. Whether any individual dies depends on many factors such as their health and strength prior to the illness, the care they receive, and so on. The percentage of people who then die is a population-level statistical fact produced by the facts of all the single cases—whether they survive or not— aggregated together.

It may be thought that it is of little consequence whether the facts of chance are determined by frequencies of occurrence or by individual propensities since such a discussion just concerns the theoretical basis of probability. However, the two theories permit a number of different possibilities and conclusions. For one thing, according to a propensity theory, the population-level mortality rate of 39.5 per cent need not tell us much, or anything, about any one individual's chance of surviving. Someone who is well cared for may have a significantly lower risk of mortality, whereas someone left completely alone may have a significantly higher risk. The figure of 39.5 per cent is just an averaging over the sum of individual risks of those with the virus. Individual risks can be aggregated and then averaged, to give us a population-level risk, but they cannot be disaggregated from population level to individual level. The frequentist does not have the resources to make this kind of distinction. The individual risk is fixed by one's membership of a population in which there is a frequency of occurrence. As a frequentist, all one might do to find a more accurate individual chance is look at a smaller sub-group, such as Ebola sufferers in hospital or within a certain age range. But this itself raises a problem for frequentism in that one individual can belong to different sub-populations, each with a different frequency of occurrence, and then assignments of probability look interest-relative; that is, relative to the choice of sub-group.

There is, though, an even more significant difference between frequency and propensity theories. Because propensity theorists insist that frequency of occurrence is only a symptom or evidence of a propensity—towards death, for instance—it is possible that the actual facts of frequency do not accurately match the real propensities. A propensity is a tendency toward some outcome, as we will describe in more detail shortly. The propensity might dispose toward a particular distribution of events, depending on its strength, but that is consistent with a number of other distributions, even if some distributions are more likely than others. The actual distribution depends on a number of other contingencies. Propensity theory is consistent with causal singularism, for instance, so it might be that there is only ever one instance of a propensity and so, regardless of its strength, it can only produce its effect or not. One cannot infer the strength of tendency from this evidence, then. Or there might just be a couple of instances, in which the strength of tendency cannot reveal itself. Or there could be some countervailing propensity that regularly prevents the first from being manifested. There is no necessity, therefore, that a propensity is correspondingly reflected in a frequency of occurrence.

Such a consequence may seem to be a disadvantage of the propensity interpretation of probability. If frequency of occurrence is a symptom of a real-world probability, instead of definitive of it, then it means that we could know all the statistical facts of occurrence but still not know what the actual strength of propensity is for causes of this type. For example, Ebola could in reality have a 42 per cent propensity to cause death but the way this was manifested in a particular population was that only 39.5 per cent died because they had a good standard of care. Propensity theory permits this possibility while simple frequentism cannot. However, it is contestable whether this is really a disadvantage of propensity theory. It might mean that the facts of probability are not as easily derived as they are under a different theory but if propensity theory is judged the better theory, on overall grounds, then one cannot reject it simply because it makes scientific knowledge more difficult. If the world is full of individual, single propensities, then we just have to deal with it. Furthermore, it may be that we have to accept a position of epistemic humility for at least some cases.

Epistemic humility is the view that there are some truths that are unknowable. It does not mean that there is nothing we can know, which is the claim of the sceptic, but that there are certain truths that cannot be known no matter how much other knowledge we have. For example, consider whether the world has infinite complexity or has some smallest possible parts that are simple, containing no smaller parts. Are electrons the simplest, smallest parts, for example, or do they have an underlying structure? We cannot know which of these two theories of the world is true, even though, it seems, one of them must be. The problem is that there is no a priori reason—using logical reasoning alone, unaided by empirical evidence—why the world cannot be infinitely complex. And there is no a priori reason why it must be. But then there is also no empirical, a posteriori proof that the world is infinitely complex or that it contains part-less simples. As Russell (1918) said, even if we found atoms that were assumed to be simple, we cannot know that they don't hide further complexity inside. There is thus no way of knowing whether the world is infinitely complex or not.

The facts of real-world probability, if a propensity theory is correct, might then similarly be unknowable in principle. If there is no reason why the facts of frequency have to reflect the exact strength of the propensity, then knowing all those facts does not tell us exactly what that propensity is. There might also be no other way of knowing that strength of propensity, in which case we have to add this kind of truth to the list of those that are unknowable. We will discuss epistemic humility again, when it comes to evidence of causation generally.

Some might say that frequentism is not the best contrast with a propensity theory. Hardly anyone still believes in frequentism, at least not naïve frequentism, as Lewis (1994) calls it. However, we saw that Lewis himself accepts a less naïve version, which he calls Humean Supervenience, and it is based on the same principle (Chapter 12). In this metaphysics, the world consists in a four-dimensional pattern of unconnected events—a history of the world from start to finish—and the truths of probability supervene or depend upon what happens in that history. Within Humean Super-venience, no judgement can be made regarding probability where there are a very low number of occurrences, for instance (Lewis 1994: 229–30). Nevertheless, this more sophisticated theory is still distinguished from propensity theory in that it asserts that the facts of chance are determined by the history of what happens. A propensity theorist says, on the contrary, that propensities produce the history of events.

20.3 What Is a Propensity?

We should now add more detail on what we take a propensity to be. This has much in common with existing propensity theories, though we will defend our own distinctive version of the theory too. Many of the elements of our theory have already been introduced earlier so the basic ideas should be familiar. We have encountered the idea of a tendency, for example, as a property of a particular that disposes toward a type of effect (Chapter 9). We also saw that to dispose toward an effect does not mean the same as necessitating it, but it does involve the possibility of producing it in some instances (Chapter 10). We should add to this the points made in Section 20.2. We can then defend a notion of a propensity as a causal property of a particular individual, process, or kind that disposes with a degree of strength toward an outcome of a certain type.

There is not much difference for us between the notion of a propensity and of a causal power, which is more familiar to some, but the idea of propensity must be of a power that is had to a certain degree or intensity. A propensity tends to issue in events—its manifestations—but it is only disposed to do so. A stronger propensity will tend to manifest more than a weaker one, though it need not necessarily do so. Two particulars can dispose toward the same outcome to different degrees, such as when two drugs have a propensity to relieve pain but one more than the other. The first drug could tend to relieve more pain than the second. When we have a number of instances of the same propensity operating, then we can get patterns of events produced, often with measurable frequencies of occurrence, which is what in Chapter 9 we called an incidence or tendential regularity: one that is less than perfect. We stated above that frequency of occurrence need not match exactly the strength of propensity because the propensity will only dispose toward a pattern of events rather than necessitate it. A fair coin will dispose toward landing heads half the times that it is tossed but there might be more or fewer than half heads over a series of trials.

We should now spend some time on what is distinctive about our account; in particular, what distinguishes it from other propensity accounts.

First, we have no commitment to a propensity ever guaranteeing an outcome. Our propensities remain irreducibly dispositional. It is worth saying this because there are some models of propensity that seem to have them collapsing into necessities in that they will produce their effects only when their chance of them doing so has reached 1 (Mellor 1971, Popper 1990). One might say that when the coin is tossed, its propensity to land head or tail is 50:50 but that the propensity gets closer and closer to 1 as the result of the toss proceeds. Just before the coin comes to rest, it might have the head clearly pointing upwards and at that point it is immensely likely to finish on a head. And the outcome is produced, on this model, at the precise point that its probability = 1. For instance, Popper says that 'Causation is just a special case of propensity: the case of a propensity equal to 1, determining demand, or force, for realization' (Popper 1990: 20).

However, we reject the principle of sufficient reason as applied to natural causal processes. A causal power—a propensity—can produce its effect even when its probability of doing so is <1. As we have argued elsewhere (Mumford and Anjum 2011: ch. 3, see also Anscombe 1971), there is no reason why causal production has to mean causal necessitation. Necessitation is just one account of how causes produce their effects and it is not a particularly convincing one. The notion of a propensity seems designed entirely for cases where a cause has a tendency toward some outcome or distribution of outcomes. If propensities resolve into necessities, it seems to undermine the motivation for invoking them in the first place, which was to allow cases of genuine probabilistic causation (see also Armstrong 1983: ch. 9 on probabilistic laws, which similarly resolves them into necessitating laws). What we offer is an account that doesn't baulk at endorsing a category of genuinely chancy causes where there is a <1 probability of an effect that nevertheless gets produced. A strong propensity would still tend to produce its effect, of course, even if there is no necessity that it will.

We also disagree with other propensity theorists over the nature of the strength— the coming in degrees—of a propensity. There are a number of options as to what constitutes the strength of a propensity. Some say it is *frequency of occurrence* (early Popper 1957, 1959, Gillies 1973), with *ratio of cases* meaning much the same (Laplace 1814, McCall 1994). This tells us that the strength of the propensity is given by the ratio of number of cases examined in which the effect occurs to total number of cases examined. And Gillies (1973) argues that the long-run frequency, if it exists, will define the strength of propensity. A different tradition relates strength of propensity to *degree of belief* (Mellor 1971, Skyrms 1977, Lewis 1980), sometimes related to rational betting on outcomes (note that understanding probability in terms of degree of belief does commit one to subjectivism, see Mellor 1971: chs 1, 2).

We have a different view from both these types of approach. Given that propensities are powers had to some degree, we need to consider this issue within the context of an understanding of powers generally. What is notable about powers is that there are some cases where there is more than enough to produce an effect (Anjum and Mumford 2018: ch. 3) and this is because strength of power cannot be measured on a bounded scale. A paperweight is used for its power to hold papers down, stopping them from blowing away. There could be a small paperweight that was minimally capable of holding down a particular pile of papers. But if you want to be absolutely sure they will not blow away, you could use a house brick for the job, or two house bricks, or more. There is no upper limit to the extent of power, in this case a downward force, that could be added; so the bounded scale used in standard probability theory is inapplicable, where all values are between 0 and 1 inclusive.

Propensities can, then, in some instances overdispose their effects. This is where there is more than enough to produce the effect. The idea should be familiar by now. There is some redundant element in the cause in that the same effect could have been produced with somewhat less, if the cause disposed to the effect with a lesser degree. This would allow us to state the extent of a propensity in terms of degree of belief, if we wanted to, in that belief itself can overdispose. As we saw in Chapter 19, there can sometimes be more than enough reason to believe something. Those cited above who account for degree of propensity in terms of degree of belief do not share our view of this, however, as they take extent of credence to be bounded. The matter is not one that we wish to pursue here, as we do not want extent of propensity to be any function of our belief states. Propensity is an ontological rather than epistemic matter. Frequency of occurrence and ratio of cases are ontological but, as we have intimated, they cannot account for overdisposing cases. You cannot have more

Figure 20.1 Asymptotic relation between extent of propensity and probability

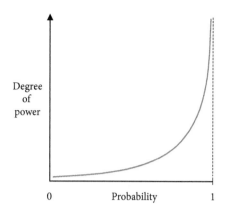

frequent than all of the time and you cannot have a ratio of positive test cases to all cases greater than 1:1. For example, the number of white swans cannot be greater than the number of swans and the number of struck billiard balls that move cannot be greater than the number of struck billiard balls. At most, those numbers can be equal.

There is a seeming problem of how propensities, with their unbounded scale of degree, relate to classical probabilities, measured on their bounded scale. This is partly a question of how to reconcile two seemingly contradictory claims within our account:

1. A propensity can sometimes have more than enough power to produce its effect.
2. A cause never necessitates or guarantees its effect.

It may be recalled that this problem was already raised in the case of beliefs, where there could be more than enough evidence to believe a theory but which, nevertheless, has no absolute guarantee of being true. These claims can indeed be reconciled on the basis that (1) concerns propensity, with its unbounded scale, whereas (2) concerns the matter of classical probability with a maximal value of 1. If one thinks of how propensity, which we now know has a non-probabilistic magnitude, relates to probability, we can see that the answer will be as given by Figure 20.1. The curve showing extent of power will be asymptotic, getting ever closer to degree =1 as the strength of a propensity increases, but never reaching 1, because a cause never necessitates its effect.

It is plausible that this kind of relationship exists between propensities and probabilities. Consider the following example. We can build some redundancy into safety mechanisms where there is a high price to their failure. Hence, we give an aeroplane two or four engines, although it can fly on only one. We also include back-up systems for every vital safety mechanism in case the initial mechanism fails. The safe completion of a flight is thus overdisposed in that there is more than enough to ensure it happens. We can entertain this thought while also accepting that the classical probability of a safe flight remains a tiny bit short of 1. And we can see from the asymptotic relation between extent of propensity and classic probability that there is a diminishing return to further overdisposing. This means that even if we spend twice as much on the plane by building in even more safety mechanisms, it would make a safe flight only very slightly more probable.

20.4 How It Comes Together: Propensities Make Chances

The distinction that is drawn between propensities and chances is useful and we can invoke it in accounting for how propensities compose to make real-world chances. The idea is that propensities and chances can vary independently of each other. For example, two people can have different degrees of propensity to be run over on the road, perhaps because one is more observant than the other. It seems possible that these propensities can be stable throughout the year. But their chances of being run over also depend, for instance, on whether they cross a road, and how often, and how busy the road is. The person with a higher propensity to be run over might, in consideration of that, cross the road only once a month and, as a result, have a lower chance of being run over than the other individual. But the chance is also fixed by the propensities of other relevant matters, such as the propensities of the drivers to run people down. If the speed limit is increased, for instance, that will also produce an increase in the chance of a fatal accident, even though the people's propensities to be run down remain stable.

This gives us, in line with Mellor (1971: 75), a picture of propensities as intrinsic properties of things, such as persons, processes, and other particulars, joining to form an overall chance. As we have argued elsewhere (Mumford and Anjum 2011: 87ff), causes can compose in non-linear ways, so we need not assume that composition of propensities is a simple matter of addition. But in spite of such complexities, the underlying idea is that individual propensities of things, disposing toward an end to some degree, can combine together in a complex to make an overall chance of a particular effect.

Further, given that we have a non-necessitating conception of propensities producing their effects, this leaves the door open for a notion of genuinely probabilistic causes. Such a model allows us to make sense of the one credible case we have of a scientifically endorsed probabilistic phenomenon, namely the radioactive decay discussed in Chapter 19. Our account gives the following interpretation. A radioactive atom has a propensity to decay with a degree stated by its half-life. There is no necessity that it decays at any time. But at some point it could. And when it does, it is not because there was any necessity of it doing so at that time. It was simply that the propensity of the atom to decay manifested itself (see also Andersen et al. 2018). This contrasts with some accounts in which the decay of the atom would have to count as uncaused because there was nothing that necessitated the decay (for instance, Heisenberg 1959: 81, 169).

There is a further important consequence. The fact that the strength of a cause— that is, the propensity—is non-probabilistically defined means that we should not follow Reichenbach (1956) and Suppes (1970) in offering probabilistic, probability-raising accounts of causation. Causation explains probability, rather than probability explaining causation. Vital for this conclusion is to show how causes do not already require prior facts of probability. In doing this, we have explained how the truths of probability can be grounded in the truths of propensity.

We accept that our account of probability based on propensities challenges some of the existing thinking on the topic but we hope the reader can see the merits in this. We will explain in Chapter 21 the impact of such a causal realist view on the way we should understand conditional probability when reasoning in science.

21

Calculating Conditional Probability?

21.1 Introducing Conditional Probabilities

Often we speak of probabilities and chances as if they were unconditional or absolute. Does this reflect their true nature? Typical chances that are referred to as unconditional are about coin tosses, roulette tables, dice rolling, and drawing of cards. When dealing with probabilities of causal claims, however, conditional reasoning seems unavoidable. We want to know how probable an effect E is given a cause C; or, if C, then how probable is E? It is therefore crucial that we understand such conditionals correctly.[1]

There are two ways we could go with conditional probabilities, philosophically speaking. One is to follow Kolmogorov (1933) in defining conditional probability as the ratio of unconditional probabilities:

The ratio formula
$P(A \mid B) = P(A \& B) / P(B)$, whenever $P(B) > 0$.

This formula is part of Kolmogorov's probability calculus, which has become the orthodoxy in probability theory. It is used in most statistical methods and is an essential element in Bayesian probability theory (Chapter 19), where conditional probability appears on both sides of the equation: $P(A \mid B) = P(B \mid A) \cdot P(A) / P(B)$.

On the ratio formula, unconditional probability is the fundamental notion from which the conditional probability can be analysed, or even defined. But there is also the opposite view, held by Popper (1959), de Finetti (1974), Gillies (2000a), and others, that conditional probability is what must be taken as primitive, and that any unconditional probability is implicitly conditional (for a detailed discussion, see Hájek 2012). The 0.5 chance of a coin landing heads is a good example of a probability that is often represented as unconditional, but even this contains tacit conditions. At least, its probability is conditional upon the coin being tossed. It is also assumed that the coin will eventually land, rather than being sucked into a tornado or left floating around on a spaceship. The surface on which the coin lands matters too,

---

[1] This chapter presents ideas developed together with Johan Arnt Myrstad.

since a flat surface allows it to fall with one side up, while a swamp, a crack, or a waterfall would not. These are just some of the conditions that fix the coin's probability of landing heads as 0.5.

We side with the camp that takes conditional probabilities as primitive. The reason for this is partly our preferred interpretation of probabilities as propensities, but there are other reasons besides that will be explained in this chapter. Since propensities are dispositional, they also have a close connection to conditionals. The irreducibly tendential nature of propensities, which we saw in Chapter 20, makes them highly sensitive to context. Different contexts tend to affect the propensity of an outcome or even trigger new propensities. In estimating chances, it is therefore crucial to concede that this estimate is conditional upon that specific context. At least for a propensity theorist.

Could the same not be said for a frequentist or a Bayesian? It seems so. A frequentist—in order to make an accurate estimate—must identify the relevant data sample from which the probability can be generated. It matters, for instance, in estimating the probability of developing thrombosis for a person whether the estimate is based on a sample of the whole population, of men over the age of 60, or of women in their early 30s, since these have different statistical averages for this particular outcome. A probability is thus conditional upon the sample from which the frequency is generated.

The same type of statistical considerations are relevant for the Bayesian. On this interpretation, a probability estimate includes both prior and posterior probabilities, where the prior probabilities are sometimes referred to as unconditional. Still, even the prior probabilities must be assigned in light of the available information, so are not entirely unconditional. Then, before assigning the posterior probability, referred to as conditional probability, one must evaluate what counts as 'relevant' new evidence among all the available data. In the case of thrombosis, if a woman is in her 30s, the Bayesian could use the statistical data to assign the prior probability. But given the new information that the woman is also pregnant, the posterior probability of thrombosis should be estimated as much higher. This could be justified through statistical frequencies and epidemiological studies, or by referring to intrinsic properties, contextual factors, and causal mechanisms.

A propensity theorist should give epistemic priority to the latter type of evidence, and consider which properties, contextual factors, and mechanisms are causally relevant in each situation. Looking at the causal mechanisms of thrombosis, we can see how this propensity is conditional upon a number of factors. Hormones during pregnancy facilitate the formation of blood clots in the deep veins, which is why pregnant women have a higher frequency of thrombosis than other women in the same age group. But it also matters whether the woman is physically active, since this promotes blood circulation, which again counteracts coagulation of the blood. If mobility is restricted, for instance due to symphysis pubis dysfunction caused by the pregnancy, the propensity of thrombosis tends to increase. Diet and general health are also relevant, since both obesity and high cholesterol affect blood flow. Age is one of the most important factors for thrombosis and most illnesses, considering that the general health decreases when one gets older. We must also take into account possible interferers of thrombosis, such as if the person uses warfarin, an anti-coagulating medication, or compression stockings, which increase blood circulation by applying pressure to the legs.

It should be clear by now that we cannot escape conditionals when estimating probabilities. Although we might sometimes speak of a probability as absolute, this does not mean that no conditions are assumed in the estimate. To have scientific tools that are suitable for dealing with probabilities conditionally is therefore crucial. In this chapter, we will explain why we think the ratio formula fails as such a tool (some of these points are developed in detail in Anjum et al. 2018).

21.2 Equation, Tool, or Phenomenon?

We have referred to the ratio formula as a tool for estimating conditional probabilities. Not everyone will be comfortable with this choice of words. The orthodox view is that 'conditional probability' simply means what is stated about the technical term 'P(A|B)' in Kolmogorov's probability calculus. Conditional probability is then defined as P(A & B)/P(B). But even though the ratio formula has become the standard interpretation of conditional probability, it is not the same as conditional probability. There are other interpretations besides, some of which treat conditional probability as primitive. Since we want to allow a different meaning of conditional probability than the technical one, it is thus useful to draw a distinction between the tool (heuristics), the results we get from using it (epistemology), and the phenomenon we are studying (ontology).

From an empiricist perspective, however, there could be good reasons for collapsing two or more of these into one. Operationalism, for instance, identifies a phenomenon (ontology) with how it is measured (heuristics). As discussed in Chapter 3, one might think of time as that which can be measured by clocks, or space as what is measured by rods. Another case of operationalism is to identify causation with what is established through randomized controlled trials. In Chapter 23, we will see why this would be a mistake. Gillies (2000b: 821) characterizes frequentism as an operationalist theory, since it defines probability as observable frequencies, which also seems to involve the identification of a phenomenon (ontology) with what we can know about it (epistemology). Hume's (1739) regularity theory of causation is an example of the latter, and so is the phenomenalism of Berkeley (1710) or Ayer (1947). A common motivation for all these approaches might be the principle of verification, which plays a central role in empiricism. This principle states that all claims that cannot be verified through observation are meaningless, including metaphysical ones.

For our purpose here, we should keep the distinction between these three. This allows us to meaningfully ask: *is the ratio formula suitable for generating accurate estimates of conditional probability?* We think the answer is no, which might be unsurprising in light of our commitment to propensity theory (cf. Humphreys' paradox, 1985) and causal singularism. Nonetheless, we think there is more to be said about exactly why the ratio formula fails as a tool for calculating conditional probabilities.

Our first concern is that conditional probabilities are here analysed in terms of unconditional probabilities. We have already indicated in Section 21.1 why we follow Popper (1959) and de Finetti (1974) in taking conditional probabilities as primitive. The source of the problem, however, lies not primarily in the understanding of probabilities, but in how conditional and causal relations are traditionally analysed. These, we will argue, need to be treated as primitive wholes rather than being analysed into something else.

Kolmogorov's ratio formula is a typical analysis, given as an equation. The equation indicates a symmetric relationship, but some asymmetry still follows. Understood as a definition, the formula must be read left to right, from the conditional to the unconditional probabilities. As a tool for calculation, however, we must use it right to left: that is, we start from the unconditional probabilities in order to calculate the conditional probability. Many equations have this asymmetry involved, even though the value on each side is the same. Take Newton's second law, $F = ma$. F might be defined as ma, but the calculation only goes one way, from mass and acceleration to force. Of course, from any two values we can calculate the third. But the case we are considering is where we know only the value of F. We then know the value of ma but we cannot, from this, decompose that value into the individual values of m and of a. And in the arithmetic equation $2 + 2 = 4$, we know that $2 + 2$ can only have one answer while 4 could be the result of many different calculations. The same could be said of chance and propensities. Even if an overall chance is made up of individual propensities, we cannot say what the individual propensities are from the overall chance. The fact that something can be calculated from certain components does not, therefore, entail that they can be analysed into those exact components. There could be composition without decomposition.

We think conditional probabilities neither compose nor decompose. They are primitive. We thus deny that they can be defined, derived, analysed, or calculated by way of unconditional probabilities. But our reason for saying this is that we think conditionals in general do not compose or decompose, as we now go on to explain.

21.3 Calculating Conditionals

Within standard propositional logic, conditional statements, 'if p then q', are represented by the material conditional, 'p ⊃ q'. But since this is defined only as a function of the antecedent 'p' and the consequent 'q', the conditional is thought to be composed

Table 21.1 The material conditional

p	q	p ⊃ q
True	True	True
True	False	False
False	True	True
False	False	True

by unconditional parts. In propositional logic, the truth values of p and q will automatically give us the truth value for the conditional, according to Table 21.1.

We see that the conditional is false only in one combination of truth values—when p is true and q is false—and is otherwise true. This might not be how we usually think of conditionals, especially not if the conditionals indicate causal hypotheses. Suppose we wonder whether you will die if you get hit by a bus. The only way to say that this is false, is if you actually get hit by a bus but don't die. In every other case, the conditional is true. So if you don't get hit by a bus, it is true that you die if hit by one. If instead you die of old age, the conditional is still true. We might not object to this, since we all agree that getting hit by a bus tends to kill you. The question is whether the truth of the conditional 'if you get hit by a bus, you die' really depends on the truth values of 'you get hit by a bus' and 'you die'. We think not: the conditional is true for other reasons entirely.

As a causal claim, we should judge a conditional statement based on an understanding of causal mechanisms and the dispositional properties involved. Whether a person would die from being hit by a bus could for instance depend on properties such as the mass, size, speed, and impact of the bus, but also on whether the person was wearing a helmet or other types of protective gear at the time of the impact.

In line with this, we argue elsewhere that dispositions are the worldly truth makers of such causal conditionals (Mumford and Anjum 2011: ch. 7). One disposition could then be used to derive a whole range of conditional claims. The fact that a wine glass is fragile means that it will tend to break if kicked while wearing boots, if struck with a hammer, if dropped from the fifth floor to the street, if thrown into a fireplace, if not properly protected during shipping, and so on. It does not matter whether we plan to actually do any of these things to the glass. But knowing about its disposition of fragility, we tend to protect the wine glass from impacts that could break it.

As a material conditional, the only thing that matters to its truth is whether the antecedent or consequent are true or false: whether you are hit by a bus and whether you die. Sometimes knowing only one of the truth values will suffice to say whether the material conditional is true. For instance, whenever the consequent is true, the conditional is also true. So since we will all eventually die of something, any conditional predicting dying as a consequent will be true. If you step on the grass, you will die. If you brush your hair, you will die. If you don't get hit by a bus, you will die. This should seem odd, even to a hard-core verificationist.

The following inferences are logically valid for the material conditional, while we think none of them are valid for causal conditionals:

1. $q \Rightarrow ((p \supset q) \mathbin{\&} (\neg p \supset q))$.
   'q' implies both 'if p then q' and 'if not-p then q'.
2. $\neg p \Rightarrow ((p \supset q) \mathbin{\&} (p \supset \neg q))$
   'not-p' implies both 'if p then q' and 'if p then not-q'.
3. $(p \mathbin{\&} q) \Rightarrow ((p \supset q) \mathbin{\&} (q \supset p))$
   'p and q' implies both 'if p then q' and 'if q then p'.

We already saw that inference 1 is problematic; that the conditional is true whenever the consequent is true. A large part of the philosophical debate over conditionals, however, is about inference 2, so-called counterfactual or subjunctive conditionals (see for instance Chisholm 1946, Goodman 1947, Stalnaker 1968, Lewis 1973a). Whenever the antecedent is false, the conditional is defined as true, which is generally acknowledged as a problem with the material conditional analysis. In the case of conditional probability, A|B, we also get a problem if the probability of B=0, since the probability of the conditional A given B could not then be estimated by the ratio formula. But this problem is avoided by the specification that the probability of the antecedent has to be > 0.

Inference 3 is rarely debated, but it is the most contestable of the three. Here, a conditional is derived directly from a conjunction of any pair of true statements. As long as p is true and q is true, we can infer that p is conditional upon q and q is conditional upon p. Causal conditionals would not allow any of these inferences, even on Hume's constant conjunction view, in which we at least need repetition and regularity. If causal conditionals were treated as material, however, we only need one instance of p and q to say that the conditional is true, at least for that case. Suppose you have a headache and want to know what caused it. Perhaps you also have the hiccups. Could you infer that the hiccups is conditional upon the headache? And vice versa? That seems highly unlikely.

Instead of using the material conditional, which after all has been subject to decades of philosophical controversy, perhaps we are better off using the widely accepted conditional probability for causal hypotheses. We could then assume that the probability of A given B can be calculated from P(A & B)/P(B). Would we then avoid some of the strange results that we got from the material conditional? It seems not. Recall that the probability of A given B is also calculated directly from the probabilities of A and B, although with a slightly more complicated procedure. In Section 21.4, we will look at some of the results we get from using the ratio formula for calculating conditional probabilities.

21.4 Calculating Conditional Probabilities

The ratio analysis of conditional probability inherits some problems from the material conditional, such as when the antecedent is false or has probability 0, plus some new problems. Here we will mention only three.

Inference 1 is similar to one we get for the material conditional; that the conditional is true whenever the consequent is true. Using the ratio formula for conditional probability, *whenever the probability of A is 1, the probability of A given B is 1 (and the probability of A given not-B is also 1)*:

1. $P(A) = 1 \Rightarrow P(A|B) = 1$, (and $P(A|\neg B) = 1$).

If applied to causal hypotheses, this would give us some strange results. For instance, if the probability of an outcome is 1, then the probability of this outcome given whatever cause should also be 1. Can we plausibly draw this conclusion?

Prima facie, it might seem strange to ever assign probability 1 to an outcome, especially if we are dispositionalists (see Section 20.3 on overdisposing). But there are cases in which this makes sense, for instance when the outcome has already happened and we want to determine the cause by looking at what happened before. Retrospective case control studies are used exactly for this purpose in medicine, where one studies a group of people who already have a condition in the attempt to find some risk factors or causes of it (Broadbent 2013: 20, 21). Here, the probability of having the condition is 1. But it does not follow from this that the probability of having the condition is 1 given whatever antecedent. What if they were never born, or got killed by a bus before they developed the condition? Or what if a cure for the condition had been found 100 years ago, so it didn't even exist today? Since the probability of A remains unaffected by whether B is true or false, the only way to make sense of this inference from a propensity perspective would be to say that the probability of A is unconditional. There are simply no conditions upon which the probability of A depends. So instead of saying that $P(A) = 1 \Rightarrow P(A|B) = 1$, we would more accurately state it as a tautology, $P(A) \Rightarrow P(A)$, and avoid the inference to a conditional. Yet, the probability calculus allows that we assign the same probability to any conditional with $A = 1$ as its consequent.

Inference 2 is similar to one we get from the material conditional whenever both the antecedent and the consequent are true; namely that the conditional is true as well. In the case of conditional probability, *whenever the probability of A and B are high, the probability of A given B is high (and so is the probability of B given A)*:

2. $P(A \& B) \sim 1 \Rightarrow P(A|B) \sim 1$, (and $P(B|A) \sim 1$).

This inference seems problematic insofar as it derives a conditional directly from a conjunction, which a propensity theorist and a dispositionalist should reject. But there could be some good reasons to accept the inference, if one is also willing to commit to certain empiricist assumptions.

Accepting that inference 2 can be used for causal reasoning, we should also accept that causation is conceptually linked to regularities, or constant conjunctions, as stated by Hume (1739). And since the conditional is estimated to have high probability only on the grounds of the conjunction having high probability, we should also commit to some form of inductivism, à la Bacon (1620). This would allow us to derive causal theories and hypotheses directly from correlation data. If we add a frequentist interpretation of probability and a truth-functional definition of conditionals, we can plausibly infer conditional relations from occurrences of atomic facts. With this move, we end up with a perfect empiricist methodology for calculating probabilities of causal hypotheses using only statistical methods and data. In other words, we have a scientific approach that fits perfectly with a Humean framework where the tool (the ratio formula) can provide accurate results (estimates of conditional probability) of the phenomenon (causation as regularities and probabilities as frequencies). We should not be surprised that this approach is less suitable for calculating propensities.

Inference 3 follows from the ratio formula and states that *whenever A and B are probabilistically independent, the probability of (A|B) will be equal to the probability of A*:

3. $(P(A \,\&\, B) = P(A) \cdot P(B)) \Rightarrow (P(A|B) = P(A))$, (and $P(B|A) = P(B)$).

What this means is that if the probability of A remains unaffected by the probability of B and vice versa (suggesting that they are conditionally independent as well as probabilistically independent), the probability of the conditional will be given by the probability of the consequent.

Again, the conditional can go either way. As long as we can assign a probability to A or to B, we can also assign a probability to a conditional having this element as its consequent. As in inference 1, this suggests that the conditional is an entirely redundant element. This fits well within a Humean or a positivist framework. If all we can know are occurrences of events, then conjunctions and negations are certainly all we need in our logic. So rather than taking conditional statements as crucial, standard propositional logic allows us to analyse 'if p then q' as a conjunction, ¬(p & ¬q): *it is not the case that both p and not-q.*

We see, then, how the estimate of conditional probability is calculated from the estimates of unconditional probabilities, just as the truth value of the conditional is directly derived from the truth values of its antecedent and consequent. Ontologically, we have argued, this has its counterpart in an empiricist framework, with causation as a relation derivable from non-causal relata, such as regularities or constant conjunctions, and causal theories as derivable from empirical data.

If we instead take dispositions as truth makers of causal conditionals, and propensities as truth makers of conditional probabilities, then causation, conditionals, and conditional probabilities must all be treated as primitive, inseparable wholes, and not analysed in terms of distinct parts.

21.5 No Conditionals in, No Conditionals out

It should be clear by now that there are no conditionals in the Kolmogorov formula for conditional probability. There is nothing in $P(A|B)$ or $P(A\&B)/P(B)$ that actually requires us to use conditional terms when we talk about it. As a purely technical term, $P(A|B)$, conditional probability does not contain any conditional words or phrases. There is no 'given', 'if', 'conditions', 'outcomes', 'effect', 'results', or even 'probability' in the formula. Mathematically, however, the formula reads: *the probability of A in conjunction with B, divided on the probability of B.* No conditionals, conditions, or outcomes are included here either.

As a tool for calculating conditional probabilities of causal hypotheses, understood in non-empirical, non-Humean terms, we must then ask where the causation or conditional probability is supposed to come from. We here side with Cartwright (1989: ch. 2), who notes that if one doesn't put causation into the theory, one won't get any causation out either. The same can be said about conditionals.

Instead of deriving causation from something non-causal, therefore, and conditionals from something unconditional, a propensity theorist should adopt a primitivist view. Causation is one of the most fundamental features of reality, making causal conditionals the most important types of statements in language. The mistake is then to try to analyse causation into something else, such as constant conjunction or counterfactual dependence. When analysing conditionals, Frege assumed the principle of compositionality: that the truth values of all molecular statements can be inferred from the truth values of their atomic constituents. He also had a similar view on causation. If we knew all the events, facts, or states of the world, this would be enough to derive all relations between them: conditional and causal ones included (Frege 1879: 134).

Why do we need to understand conditional probabilities? And why use the ratio formula in our scientific methods? It seems that the only reason is that we really need a tool for studying causal connections in the world; whether one factor influences or affects another. Note that both 'influence' and 'affect' are causal notions, suggesting that the risk or prospect of the outcome should, in some way at least, be conditional upon something else. It seems, then, that we would not be using the ratio analysis in our scientific work unless we had some interest in causation. But, as we have shown, it is far from a perfect tool for this purpose.

PART VII

External Validity

22

Risky Predictions

22.1 Explanation's Poor Relation?

Part VII of this book is concerned with prediction, particularly as informed by experimental and other knowledge. This puts a focus on the external validity of experimental results and the way in which such evidence is a basis for prediction. We can perform carefully managed experiments, such as randomized controlled trials (RCTs), but what do they tell us about what will happen elsewhere, in less controlled circumstances?

It is easy to overlook the topic of prediction, even in philosophy of science. As Broadbent (2013: 84) noted, the topic has received remarkably little attention. One might assume, for instance, that the main task of science is to discover and explain causal connections. But while these are certainly important tasks, the reason they are important is that such causal knowledge can be used as a basis for prediction: and this is the thing that really matters. Of course, once one has an account of explanation, it might be thought that an account of prediction follows pretty easily. Explanation and prediction could be two sides of the same coin and thus amenable to a unified account. Perhaps the difference between the two is small, coming down simply to what takes the place of explanandum, as in the Hypothetico-Deductive model (Hempel 1965, Popper 1972, also Cartwright 2007a: 25, 26). From the hypothesis under consideration, together with auxiliary hypotheses, you can draw empirical consequences, which then constitute the predictions of the hypothesis. Broadbent (2013: 103) says we can think of explanation as inference from effect to cause and prediction as inference from cause to effect. One might also think that prediction is adequately covered by topics such as counterfactual reasoning, induction, and laws of nature. We agree with Broadbent that it is not adequate to treat prediction as a footnote to any of these topics.

Irrespective of the details of the explanatory model, there is nevertheless a significant metaphysical difference between explanation and prediction. When we explain something, it has already happened and is thus taken as given. A prediction is often (admittedly not always) of something that has not yet happened even though it might be expected to do so. Explanation involves applying theory to what has already occurred. What is the point of explanation (other than intellectual curiosity), though, if we don't use what we learn from it to predict future cases?

Some of these predictions are vital; indeed, they can be matters of life or death. We want to know the:

- future extent of climate change;
- risks and benefits of nuclear energy;
- effect on UK economy from leaving the European Union;
- effects of medical interventions;
- projected biodiversity trends and their influence on the global eco-system.

Prediction can also be applied to resolve specific questions:

- After how long will radioactive waste become safe?
- What will be the health improvement of a specific dietary alteration?

One might think that prediction doesn't require or even permit a deep philosophical analysis because it is such a basic idea. But we will see that there are outstanding issues about how predictions work and what we can expect from them. This chapter will consider those bigger issues. In Chapter 23 we concentrate more specifically on the external validity of RCTs, which for many represent the gold standard of evidence-based investigation.

22.2 Necessitation and Fallibility

There are many predictive successes, which we should acknowledge. Some predictions are very reliable. We are able to predict eclipses of the sun to the second, for instance. And even in cases where there is not quite that exact accuracy, we have enough of a predictive grasp of the world that we can act on the basis of expectations that are regularly fulfilled. It will be good, therefore, if we have a model of prediction that explains how it can be so successful. There are attempts to do this. The aforementioned Hypothetico-Deductive model, for instance, purports to deliver certainty in prediction, given a hypothesis and subsidiary assumption.

But it must be admitted that predictions are also fallible. We sometimes make a prediction based on the best information available but see it disappointed. There can be experimental failure where some expected result is not realized. We know that things can go wrong in any process in the natural world. Indeed, scientific progress can occur when something surprising happens that confounds our expectation. This point is worthy of note because, as well as explaining what happens when a prediction goes right, we also need to understand what happens when a prediction goes wrong. As the name suggests, the Hypothetico-Deductive model indicates that as long as you reason correctly and your hypothesis is right, then the prediction follows as a matter of deductive certainty. Of course this is not the main point of the model of prediction, which is to use the predicted outcome as either confirmation or falsification of the hypothesis. But the model only gives us that on the grounds that the prediction is implied in the hypothesis and is thus deducible from it. It is this latter point that we wish to challenge here. Predictions can be—indeed are—fallible even when they are derived appropriately from a true theory. How can we explain that?

We have already encountered the necessitarian conception of nature, which claims that what happens does so as a matter of necessity. If one reasons soundly from a true theory, then, any such prediction that is produced must be true. This would give us a necessitarian theory of prediction. We urge, on the contrary, that predictions are essentially fallible and any model that does not allow and explain this is inadequate. This is so even for 'good' predictions, that is, predictions made in the right way adopting appropriate methods. A prediction produced like this can be good even if its predicted outcome does not occur.

Various considerations challenge the necessitarian view of prediction. We will outline these before offering our own tendency-based positive account of prediction. We claim it is consistent with both the fallibility of prediction and its success in many cases. We will then consider where this leaves the external validity of scientific results, which is of vital importance to the question of predictive success.

The first challenge to the necessitarian view is plain fallibility. By this, we mean that it must be taken as a datum that predictions often fail, even when based on the best available knowledge. Despite advances in seismology, we still cannot predict the time and place of earthquakes to any useful degree. We sometimes predict the wrong flu virus so the jabs we give to people for the winter have no effect on what emerges as the dominant strain, as happened in the USA in 2014/15. This led to vulnerable groups being inoculated against the wrong strand based on what looked like being dominant nine months earlier. In the time the vaccine was manufactured and distributed, the flu virus mutated in an unexpected way. Fallibility is thus just the empirical fact that predictions sometimes fail, even if they are based on good evidence, and a necessitarian owes us an explanation of this.

Of course, a necessitarian is at liberty to claim that in cases of predictive failure, we were reasoning from false hypotheses, rather than it being anything to do with the world itself. But there is another challenge to consider. Some phenomena are regarded as irreducibly indeterministic, for instance within some interpretations of quantum mechanics (see for instance Bohr 1937, 1938, 1948, von Neumann 1955, Heisenberg 1959, Feynman 1967, Healey 1992). This raises the possibility that ours is an indeterministic world. If that is so, then there are at least some phenomena that cannot be predicted with certainty. This does not mean that unpredictability is definitive of indeterminism, as we will explain. Instead, we say that worldly inde-terminacy means that some outcomes are not a matter of necessity and, for that reason, are not calculable from a set of initial conditions. Indeterminacy is a basis for unpredictability.

However, as is now well understood, even a deterministic system could be unpre-dictable in principle, if it is chaotic (Lorenz 1963). The existence of chaotic phenom-ena would thus be another explanation of a lack of necessity for our predictions. A chaotic system is hypersensitive in the sense that even the slightest difference in initial conditions can lead to a large difference in outcome. Any prediction that we make, on the assumption of initial conditions, has to be based on a specification of the variables of the conditions to a finite number of decimal places. We could then have two systems with exactly the same values for its variables if specified to four decimal places. However, these two systems could evolve (deterministically) very differently if those values differ at their fifth decimal place. This makes the system unpredictable in principle because the argument applies for any number of decimal places.

The next point against necessity is related to this. It might be urged that in order to be entirely reliable, predictions should be based on exact values for the initial conditions. But the foregoing point within chaos theory shows that this is not possible. The problem is that 'exact' is an imprecise term. To how many decimal places must you specify it in order for a value to be exact? And if you are to start an experiment on two fields of corn of exactly the same height, what does that mean (Sober 1999, see Cartwright 2007: 71)?

Finally, we also have to accept the force of Hume's arguments against necessity. We do not endorse Hume's philosophy. But his argument against necessity is one that still holds power. If A necessitated B, then it should be impossible to have A without B. But we know that it is possible for A to be the typical cause of B, where some instances of A can nevertheless occur without B. We have used this kind of argument ourselves, when we discussed additive interferers (Chapter 8). The argument remains effective against a necessitarian view of nature and of prediction in science.

We do not think the world works in the way Hume described. If anything could follow anything, we would have no rational basis at all for prediction. Hume (1739: I, iv, 7) reluctantly swallowed this conclusion, accepting that inductive inference is not rationally grounded. But we do not think the world works by necessity either. If it did, we could have predictions that were in principle certain. We will therefore go on to offer an account of prediction in which good predictions will be useful despite being fallible and have a tendency to be true but without any guarantee.

22.3 Predicting with Tendencies

In our account, predictions are based on knowledge of tendencies, which we think may usefully be represented in a vector model (Figure 22.1). Once a resultant vector, R, is calculated, it gives us an indication of the direction in which the overall situation disposes.

It is characteristic of our account that this does not necessitate what happens: it only shows us what will tend to happen, where tendency carries an irreducibly dispositional force that is stronger than pure contingency but weaker than necessity (Chapter 10). Knowing that objects gravitationally attract in accordance with the gravitation law, for instance, tells us that that they tend toward each other. The more

Figure 22.1 A vector with the resultant
tendency R

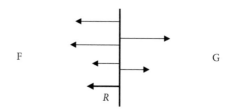

F

G

R

massive the object, the more others will tend to move towards it. This presents a very solid basis for prediction of the movement of objects. As Cartwright (1999: 27) has pointed out, though, it doesn't tell you where a dollar bill will land if taken by the wind in a busy city square. While the tendency to fall towards the ground is still there, all sorts of other tendencies will be acting on it, such as wind direction and whether the bill is folded, which can counteract the gravitational attraction. It is quite conceivable that someone catches the bill and puts it in her pocket.

The more one knows about the powers at work, operating on a particular situation, the better one will be able to predict the eventual outcome. The basics of this theory thus equip us to explain how predictions can often be reliable to an extent even though they are also fallible. Knowing that a cause tends toward a particular effect does not offer a guarantee; but it says more than that the effect is a mere possibility. The dispositional account also explains how the *ceteris paribus* clause that is often attached to predictions can be non-vacuously true. To say that shaken gelignite explodes, *ceteris paribus*, means that it has a real disposition to do so, even though there are circumstances in which it might not.

In many cases of predictive failure, the reason for it could indeed be that the prediction was based on inaccurate or incomplete explanation. One might be ignorant of some of the powers that were at work. Doctors did not wash hands between patients because they did not understand that there were bacteria that had a causal power to spread diseases. Such a power could not, then, enter into medical considerations and doctors were unable to predict that washing hands would lead to a reduction in mortality rate. In another type of case, one might not understand that a substance is a non-Newtonian fluid and then be unable to predict that it can harden when struck, like cornflower does (Waitukaitis and Jaeger 2012).

A different type of error is possible even in a case where all the effective causal powers are known. One might know all the individual powers operating on an effect but not understand how the powers combine, compose, or work together. Not all composition is linear. Powers sometimes compose in non-linear and even emergent ways. An example of non-linearity is where two drugs individually dispose toward a certain outcome but when taken in combination dispose in the opposite direction. This happens with the case of clonidine and sotalol, which individually can lower blood pressure but in combination can raise blood pressure (Saarimaa 1976). Emergent composition is where a combination of powers, through their causal interaction and transformation, produce a new higher-level causal power: new in the sense that it is not had by any of the component powers prior to them entering into the related whole (see Chapter 14). Hence, it may not have been predicted that certain combinations of common elements could combine to form a living organism.

However, it is also distinctive of our account that it allows predictive failure not just to be a result of ignorance. If the world itself works in a tendential way, then even with perfect knowledge of the powers of things and the ways in which they compose, there is no guarantee that predictions must come true. Such thinking is no longer entirely alien, given that we have come to understand that at least some parts of the world could operate in an indeterministic way. Tendencies do not reduce merely to indeterminacies, on our view, but the notions of causal production and causal necessitation should be kept separate (Anscombe 1971). A cause can tend toward an outcome and sometimes succeed in producing it, where that cause can be a propensity, having a strength of tendency towards the outcome.

The tendency account explains a number of other features we find in prediction. For example, Broadbent (2013: 89, 90) rightly distinguishes the product and activity of prediction. Prediction as a product can be thought of as a statement of the thing predicted, such as that it will rain tomorrow, which will be either true or false. The activity of prediction is the processes one goes through in order to deliver such a product. 'Predict' is also a verb and what is vital about this distinction is that it allows the possibility of a good prediction—one based on all the best available evidence and reasoned correctly—which nevertheless turns out to be false. Recall in Chapter 3 that we distinguished two types of weather prediction. One was scientifically based using all the meteorological evidence, the other was made by a supposed psychic. It is possible that the meteorologist gets the prediction wrong but that it was still a good prediction. Similarly, if you follow a hunch and predict the dice will land on a six, your prediction was not good, in the activity sense, even if it turned out true. Following hunches is not regarded as a good, scientific method of prediction. Of course, if a 'method' such as following hunches was found to get predictions right more than a scientifically validated method, we may well start to question whether there is something in it after all.

A good predictive method, though it need not deliver truth, must at least aim at truth. And, as results matter, a good method should tend to deliver truth, otherwise we would rightly question whether it is good. But predicting the truth is not all that makes it a good prediction. Indeed, the truth of the product of prediction is neither a necessary nor sufficient condition of the prediction being good.

Some points made in relation to Popper's falsificationism explain this. It is easy to make a true prediction: for instance, that the sun will rise tomorrow. A good prediction, however, should also be useful in the sense of being otherwise unknown or unexpected. The prediction that on a particular October day in Oslo it will be 30°C will, if true, be a very good prediction. It is a prediction of something that has a low prior probability, given all our background knowledge of the climate in Oslo. And it will allow us to revise our behaviour accordingly, for instance, wearing shorts. The prediction need not even be strictly true in order to retain those benefits. If the temperature only reaches 29°C, many of those revisions in behaviour will still pay off. This shows, to adapt part of Popper's view, that a very specific prediction can be of more use than a very broad one even if the former is false while the latter is true.

In accordance with this point, we can take it as a norm of science that a good prediction activity will tend to get it right. What counts as getting it right is clearly more complicated than the prediction simply being true, as we have just seen. The prediction is better when it is specific and the benefits of such specificity may survive even if the prediction is strictly false. Again, our notion of tendency, as in tending to be right, is not reducible to simple quantitative measures, hence we cannot say that such predictions will *usually* be right (cf. Broadbent 2013: 93) as that imposes a quantitative measure on to the notion of tendency. There may be some instances where a predictive practice is warranted even if it gets it right slightly less often than it gets it wrong. A method of predicting the time and place of earthquakes would still be immensely valuable even if it 'only' got it right 45 per cent of the time.

A further clarification can be made. A prediction need not always be of something that hasn't yet happened. Suppose a lorry of sheep arrives at a farm and you hear that they had a long journey without being watered or fed. You might predict grimly that around 30 per cent of the sheep will be dead. What is *predicted* is the obtaining of a certain result when the lorry is opened up. But the deaths themselves will have already occurred (or not) before we look. Hence, it is an observation that is being predicted in this case, rather than the facts being observed. But it does show that prediction is essentially about something unknown that may come to be known, even if most cases of this type are about things that have not yet occurred. Similarly, one could predict that digging in a certain location will uncover dinosaur fossils but the fossils are of course already there before any digging starts.

Basing predictions on knowledge of causal powers does not necessarily mean knowing all the mechanisms of those powers. As Broadbent (2013: 111) explains, Lind predicted that lemons and oranges would cure scurvy but he didn't know the mechanism (vitamin C). Similarly, many had confidence that chocolate was a cause of migraine though we are only just starting to understand the mechanism (Lipton et al. 2014). One may, of course, believe that there is some mechanism linking chocolate and migraine but one's belief in the causal connection may still be confident even without knowledge of the detail, which for many practical purposes is irrelevant. What counts for the sufferer is that if they stop eating chocolate, the migraines will go away. They can treat the mechanism as a 'black box'.

A final matter on our positive proposal for a causal powers account of prediction concerns probability assessments. An estimate of the probability of something occurring is not a prediction, as saying that there is a 60 per cent chance of rain is consistent both with it raining and with it not raining. It does not commit to anything happening: including any particular distribution or frequency of occurrence. And one might argue the same of tendency claims, of course, if one says that there is a disposition or tendency towards rain tomorrow. As Fuller and Flores (2015: 50) suggest, however, this could technically count as a predictive practice as it is an expression of a probability of (or tendency toward) a predicted outcome. And such an assessment being consistent with any outcome or distribution in a single test situation does not, of course, mean that there are no rational grounds for taking such a prediction to be false in the light of the outcomes. One might revise a probability assessment, for example, that persistently over time is not born out by the facts.

22.4 Prediction and External Validity

We have yet to draw the connection between experimental results and prediction. It is clear that we would say that the value of experimentation for prediction is that it gives us knowledge of the causal powers of things and their interaction, which can then be applied outside of the experimental context. A good experiment will reveal the causal powers of the subjects, even if it needs artificial circumstances in order to do so. Experimentation and other forms of evidence thus make for better-informed predictions.

However, as is well known, there are problems around the external validity of such evidence and these problems also add to the fallibility of prediction. The problem of external validity is simply the point that what is found in one experiment, study, or set of circumstances might not apply to a different set of circumstances. There can be questions of internal validity—whether the data support a causal hypothesis, for instance—while external validity concerns whether that same causal hypothesis is applicable outside the studied cohort. Again, this could be seen as a problem of inferring from what happened in experiments to what happened or will happen in reality. The general problem has a number of specific sub-types, however, depending on what sort of inference one is drawing. The first two of these come from Fuller and Flores (2015) to which we add another case.

1. Generalization (extrapolation, transportation): from sample (study population) to general (target) population.
2. Particularization: from group level (target population) to individual (within that population).
3. Replicability: from one group or sample to another.

All of these inferences contain pitfalls, largely around knowing whether the experimental group is sufficiently like the target group. In the case of 1, for instance, there are problems concerning the differences between the experimental population and the general population. A drug may be tested on mice, for instance, which are bred to be genetically identical and have virtually no immune system as they are raised in a sterile environment. How confident can we be in results from such a study when they are applied elsewhere? Case 2 raises the problem of the ecological fallacy. It is unreliable to infer from a group to an individual, especially where that concerns statistical matters. There will be an average height for a group, but plainly that does not allow an inference to any one individual's height. This is not to deny that population-level predictions can be useful: it is still useful to have a predicted incidence of heart disease in a population, for healthcare planning purposes, even if one doesn't know which individuals will be affected (Fuller and Flores 2015: 52). Case 3 concerns the matter of a result in one group carrying over to another. There are many cases of failures of this kind (Chapter 28), for instance the nutritional education programme that worked in Tamil Nadu but failed in Bangladesh because of the different social circumstances there (Cartwright 2012).

The advantage of the laboratory is careful control of the factors, their measurement, and recording. There are similar advantages from forming models that allow us to explain some complex phenomena, sometimes isolating the causal powers involved. Prediction brings those complications back into play, however. As Cartwright (2007a: 29) says, the problem is how we know that real-world situations are like those of the model, experiment, or study population. Broadbent (2013: 107) adds that extrapolation alone is useless. What does the main work of prediction is the knowledge of relevant similarity between the known cases and cases under consideration. This, we have argued, comes down to knowing which causal powers are at work and what outcomes they tend to produce.

23

What RCTs Do Not Show

23.1 Evidence-Based Decision Making

Earlier (Chapters 11 and 12) we discussed the role of data for causal theories, arguing that such theories cannot be directly derived from data. Instead, it seems, data are at least to some degree dependent on theories and vice versa. In Chapter 22 we saw that causal predictions are essentially fallible, even if we could have access to all the relevant data. It is now time to take this discussion to a more practical level. Evidence-based decision making is a scientific approach in which causal conclusions and predictions are thought to follow directly from data. It first started with the introduction of evidence-based medicine back in 1992 (Guyatt et al. 1992) but the evidence-based approach has in recent years expanded into other disciplines, including public policy. That policy decisions should be evidence based has now become a matter of national and international government policy (see e.g. Haynes et al. 2012, Cairney 2016).

That decisions ought to be based on evidence might seem uncontroversial as a norm. As discussed in Chapter 22, a prediction must be understood as conditional upon available data and theoretical knowledge. Any prediction, and its subsequent decision, must thus be informed by what we already know. The matter of controversy regarding evidence-based decision making, then, is rather the *type* of evidence on which it should be based. Not any kind of scientific results count as evidence in this context, but preferably those results that are produced by performing certain types of epidemiological studies; so-called randomized controlled trials (Howick 2011: 5).

An RCT is an experiment, insofar as it tests the effect of an intervention, controlling for all other factors that could causally affect the outcome. This is done by randomly and blindly assigning trial participants to two groups. One group will get the tested intervention (the test group) and one group won't (the control group). The control group could get a placebo or a standard intervention, or they might get no intervention at all. The importance of the randomization is that it is supposed to ensure that any difference in outcome between the two groups is entirely due to the intervention, and not to chance or any other factors, which should be evenly distributed (for details on RCTs as a method, see Broadbent 2013, Howick 2011, Cartwright and Hardie 2012).

RCTs are considered best suited for guiding policy decisions and are referred to as the gold standard within the evidence-based framework (Sackett and Rosenberg 1995: 408, 409, Sackett et al. 1996: 72, Worrall 2002). In social, economic, and biomedical sciences, RCTs and systematic reviews of RCTs are ranked the highest forms of causal evidence by policy-vetting agencies such as GRADE—the Grading of Recommendations Assessment, Development and Evaluation Working Group (Balshem et al. 2011). This means that if we want to know whether an intervention or exposure causes a certain outcome, RCT is thought to be the best method to decide. Today, results from RCTs are commonly used to guide expert advice within a wide range of areas, including medicine, management, education, economics, and social policy (Deaton and Cartwright 2018).

According to Cartwright and Hardie (2012: 123), there are some clear advantages of RCTs for discovering causes. One is that they are *self-validating*, meaning that reliable results are guaranteed by the study design. That the results are reliable must here mean that the results we get from RCTs are exactly the type of results we are interested in getting. In this context, one is interested in finding out whether an intervention caused an increase in the outcome within the study sample. We have some concerns about drawing causal conclusions from an RCT, which involves a commitment to the contrastive theory of causation (cf. the discussion in Section 15.4). But another advantage, emphasized by Cartwright and Hardie, is that no prior mechanistic or other background knowledge is needed to generate a causal conclusion, since all other potentially relevant causal factors should be distributed evenly between the two groups. Ideally and in theory, therefore, RCTs are generally thought to have strong internal validity (see Cartwright 2007b on 'ideal' RCTs). More controversial, however, is the question of external validity of RCTs; whether the results from the study can be used to make predictions and inform policy in other contexts. Cartwright and Hardie (2012) refer to this problem as the inference gap from 'it worked there' to 'it will work here' (see also Cartwright 2012).

External validity is not the only concern one could have about RCTs as a method. In Chapter 15, we presented some philosophical problems with the way in which RCTs identify causation via difference-making. In this chapter, these arguments will not be repeated, but they are nevertheless relevant for evaluating the scope and limitations of RCTs. Instead, we will explain why the results that we get from RCTs systematically fail to take into account certain types of causally important knowledge. When allowing RCTs to inform policy decisions, it is therefore essential that we are aware of those elements that are excluded from the studies. Otherwise, our decisions will be based, not on the best available evidence, but on some very specific and artificial form of evidence. What it means to base a decision on the best available evidence thus depends on what is meant by 'best', 'available', and 'evidence', as we now go on to discuss.

23.2 Causally Relevant Elements Excluded from RCTs

There are some things that RCTs do not show, but which should be considered when making decisions using evidence from them. Our main purpose in this chapter is simply to explain why we cannot trust RCTs to offer the full causal story. If so, they are not sufficient for making a fully informed decision. We will then go on to argue that evidence provided by RCTs, or any other method, could never alone tell us what to do. Other and more detailed criticisms of RCTs can be found in Broadbent (2013), Cartwright (2007b, 2010, 2011), Howick (2011), Worrall (2002, 2007, 2010), Marchal et al. (2013), and Rothwell (2005).

23.2.1 Severe effects

RCTs are thought to be the best method for generating causal evidence, but their scope is fairly limited. Not all types of effects are suitable for being tested through RCTs. One cannot, for instance, use this method to settle whether driving while angry increases traffic accidents, whether experiences of sexual abuse in childhood causes chronic illness as adults, whether systematic embezzlement is linked to company bankruptcy, or whether being hit in the head causes brain damage. If one performed RCTs for these interventions, participants risk serious harm or even death as a direct result of the trial. Whenever an intervention poses a severe risk on the study participants, therefore, one would not run an RCT to test it. A *reductio ad absurdum* of this norm is made in the satiric research paper 'Parachute use to prevent death and major trauma related to gravitational challenge: Systematic review of randomised controlled trials' (Smith and Pell 2003: 1459). Here, the authors argue that we cannot advise the use of parachutes to prevent death from high falls, since all we have to go on in this case is anecdotal evidence. The example illustrates why RCTs should not be a scientific requirement for establishing causation.

That evidence of causation in cases of severe effects cannot be provided by RCTs shows that we have to widen the scope of evidence. Much of our causal knowledge is useful exactly because it can help us decide also what *not* to do. Although some of these effects are known through experience, they will have to be discovered from sources other than RCTs. We cannot say absolutely that this experience is automatically less valuable as evidence because it wasn't generated by one particular method. When choosing between two types of intervention, therefore, we should be allowed to consider *all available* evidence, and not be restricted by what we can get from RCTs.

23.2.2 Risk groups

Results from RCT studies are thought to carry over to the relevant population because they are obtained from a representative sample. But are they? Just as we cannot perform RCTs to test for severe effects, we cannot include participants in the study if they are at risk of getting a negative effect from the intervention. For ethical reasons, certain groups must therefore be excluded from participating. If someone has allergies, or is mentally unstable, for instance, then certain interventions could put them at risk. Some groups are generally thought to be more vulnerable to adverse effects. In medicine, typical risk groups would include children, older people (Scott and Guyatt 2010), and people with multiple chronic conditions and complex medical needs (van Spall et al. 2007, Fuller 2013). Pregnant women are also treated as a high-risk group since the discovery that the drug thalidomide caused thousands of babies to be born with malformed limbs in the 1960s (Kim and Scialli 2011). As a result, hardly any medications are tested on or prescribed to pregnant women today (Foulkes et al. 2011, Shields and Lyerly 2013).

The strict inclusion and exclusion criteria for RCTs means that, for some trials, a very small proportion of the relevant population is eligible. Rothwell (2005: 85) reports that exclusion rates can be over 90 per cent already at the first stage of selecting participants, and that it becomes even higher during pre-randomization run-in periods. If someone shows an adverse effect to the intervention, they could be excluded from the trial. In one RCT, performed by Gómez-Marín et al. (1991) to study the effects of salt on blood pressure in children, 93 per cent of the participants were excluded in the run-in period (Rothwell 2005: 86).

When anyone who is at risk of adverse effect is excluded from a trial, the results from RCTs fail to accurately represent its target population. Reports of adverse effects will then come in spite of attempts to avoid them, meaning that the intervention will tend to appear safer in light of the evidence than it actually is. If we then go on to assess risk post-trial, by monitoring the results of the intervention in individual contexts, we might discover outcomes that were not found in the trial. It seems, then, that any risk involved in an intervention cannot really be studied within an RCT. Instead, they must be discovered after the official trial is over, probably in those belonging to the risk groups, and through methods that are generally thought to provide weaker evidence of causation than an RCT. The ethical implications of this seem more severe than the epistemological and ontological ones, but for the purpose of this book we concentrate our focus on the latter two.

23.2.3 Variations

RCTs typically test for common effects and causes within a group. This means that heterogeneity or variations must be ignored or followed up separately. But one type of intervention might not have the same type of effect in all contexts. There could be causally relevant differences. In Part II, we saw how the classical Humean notion of causation assumes that the same cause should always give the same effect, and that this forms the basis of some scientific methodology. RCTs are no exception in this connection. Causation is thought to be that which brings about the same type of effect over a variety of contexts, what we called robustness. Woodward (2003) even thinks of causation as a relation of invariance.

Although RCTs include variations in their study design, this is not what the test is designed to show. On the contrary. Variations and randomization are considered important parts of RCTs only to ensure that all causally relevant differences are represented and evenly distributed in the samples. It is not because one wants to study the actual differences in outcomes. One could, of course, use RCTs to study a particular type of variation by singling out a sub-group, for instance Montessori schools, low-income families, or children with type-II diabetes. But within each such sub-group, homogeneity—not heterogeneity—is what one is studying. As a result, any variation in sample and outcomes must be treated as noise: not predicted by the hypothesis.

Could we take care of variations through metastudies of RCTs, in which the outcomes of relevantly similar studies are synthesized and compared? This is an ongoing debate in the medical sciences (e.g. Eysenck 1994, Steurer et al. 2014, Buetow 2015, Miles 2015). Obvious advantages of meta-analyses are related to the amount of data. The idea is that when we have more data, we get a more representative sample, thus stronger external validity. On the other hand, if each individual study reports a single outcome for a single intervention, a metastudy will have to summarize these results into even more general conclusions. By doing this, one might end up comparing contexts, samples, scopes, and results that are not naturally comparable (Eysenck 1995: 31). If the aim is to account for variations, therefore, metastudies seem to do exactly the opposite.

23.2.4 Individual propensities

Recall how the norm that decisions should be based on 'evidence' largely means the outcomes of epidemiological studies, preferably from RCTs. We can say that the results should be used to *inform* or *guide* our decisions. How would they do so? Epidemiological studies are typically made on a group level, using statistical methods. But, in medicine at least, where the evidence-based approach began, decisions are made on the level of the individual. For education policies, one might use data from a sample of schools to make decisions about class size, teaching method, or management strategy elsewhere. Whenever we go from the level of groups or populations to an individual case or person in this way, we make an inference from statistical frequencies to individual propensities.

In Part VI, we explained why frequencies and propensities should be kept conceptually and ontologically separate. Propensities, unlike frequencies, are qualitative and intrinsic rather than quantitative and proportional. If a teaching method is shown statistically to improve the students' results in one population, this is a purely quantitative measure about the frequency of an outcome within this sample. But it is perfectly possible that individual students, also in the tested sample, get a lower grade with this method or even no change. All that can be shown with a population study is that the overall outcome improves. It cannot show why some students' results improve or why others' drop. That is a theoretical question, which falls outside the scope of RCTs. Which method works best could depend on the class size, the knowledge of the teacher, or on the individual personality and abilities of the students. Perhaps some students concentrate better on their own, while others need more personal follow-up.

In medicine, there are risks involved in applying the results from RCTs directly to individual patients (Greenhalgh et al. 2004). The intervention that works for most patients might not be suitable for all. Unless we assume that each individual patient is statistically average, which is unlikely, individual propensities cannot be inferred from statistical frequencies. This is a well-known problem, also referred to as the ecological fallacy. A question remains whether the fallacy could be avoided by singling out smaller sub-groups, thus taking into account more of the causally relevant factors for the individual. But this seems problematic, since one would still have to make a decision based on the averages within these smaller groups. If we keep finding more and more relevant sub-groups, by including more specifications, we eventually end up with a sample of one individual. In some cases, this is the only sample used to study an intervention, as in N-of-1 studies, where only the one patient is involved.

This is analogous to Russell's argument against using a notion of total cause to save causal necessitation (see Section 8.4). If one keeps including more elements in the cause to make it fit a certain context, it is less likely that the cause will ever happen again. Instead, one ends up with each cause being very large, and effectively unique. As causal singularists, we have no problem with causal uniqueness. But since epidemiology as a methodology is closer to Hume's regularity theory, no possibility of repetition means no possibility of obtaining causal evidence, by its own standards.

23.2.5 Biases

There are a number of other, perhaps more serious challenges for the external validity of RCTs than we have discussed here. Biases of selection and publication are often discussed, but they are not specific to RCTs.

If a study shows no effect of an intervention or the results are inconclusive, they might never be published, leaving the results of the study unknown. Meta-analyses are thus in danger of showing an overly optimistic trend: that the intervention has a positive effect. This is why there is an increasing pressure on scientists to publish negative results and some journals specialize in this (for instance, *Journal of Negative Results in Biomedicine, Journal of Negative Results—Ecology and Evolutionary Biology*, and *Journal of Articles in Support of the Null Hypothesis*).

Selection bias is also a general problem. The external validity of a study depends on the sample being representative of the general population, including economic, social, and ethnic diversity. But this is often restricted by the local source (Fuller 2013: 645). For instance, many studies within economics and psychology use undergraduate students from American academic institutions as their sample, since they are easily available test subjects. Typical for these, however, is that they come from so-called WEIRD societies: Western, Educated, Industrialized, Rich, and Democratic (Henrich et al. 2010). So how representative are they?

Other challenges are particular to RCTs, such as if there is no available intervention or no available placebo. It is difficult to find causally neutral placebos, that is, placebos that have no effect on the outcome (Howick 2011: 80ff, Broadbent 2013: 23). This might lead to a different sort of bias, namely which types of interventions are suitable for being tested through an RCT. The health benefits of pharmaceutical interventions will for instance be easier to test and control than social or psychological factors, such as personal relationships and social status. Race, gender, income, or sexual orientation are known to be causally relevant in many contexts, but RCTs cannot be used to demonstrate this. Furthermore, being in a loving relationship might have a huge health benefit, but how could we offer real love to an intervention group? And what could possibly count as placebo love? Trial drugs seem to face no such difficulty, leading to a suspicion that the method of RCT structurally favours pharmaceutical solutions.

23.3 Evidence-Based Policy—a Word of Caution

According to Cartwright and Hardie (2012: 123), RCTs 'clinch', or settle, causal conclusions in the sample studied, in a deductive manner. They have strong internal validity. What they cannot do, however, is tell us what will work elsewhere. They lack external validity. We agree with their second conclusion.

In this chapter we have seen that a number of causally relevant elements are excluded from the results of RCTs: severe outcomes, risk groups, variations, individual propensities, and biases of publication, sample selection, and type of intervention. The fact that RCTs are unsuited to comprise these elements makes it even more important to take them into account when we interpret the results from RCTs and use them in decision making. When doing this, we should think of everything that was taken out in the study. If we fail to consider what is excluded already at the outset of an RCT, we risk drawing conclusions that are not warranted. In contrast, being open and explicit about what is excluded from an RCT allows a more realistic interpretation of the results. This should also make us more cautious about applying these results universally and unconditionally (Hjelmesaeth 2014). But to what extent does evidence-based policy allow such caution?

The norm that policy should be based on the best available evidence seems irrefutable. But we should not expect that a *decision* can be made just by considering the evidence. The best possible RCTs and meta-analyses *might* show which of the known interventions benefits most people. But there is no policy that automatically follows from that. For this we need another norm, derived not from the realm of science but from ethics.

We have argued elsewhere (Anjum and Mumford 2017c) that evidence-based policy involves a normative element: of rule utilitarianism. By recommending the intervention that has been shown to benefit the most, one effectively accepts the utilitarian principle of generally applying the rule that maximizes utility. But rule utilitarianism has been criticized and for good reasons. If one really wants to maximize utility, one should be allowed to break the rules in cases where doing so produces more benefit (Smart 1973, Williams 1972: 102). And since few RCTs show a ubiquitous benefit from an intervention, there will always be some people for whom no benefit is produced, or even harm instead. For this, knowledge about causal mechanisms and qualitative data about the local context of application seems more useful than quantitative information about what works elsewhere (for a discussion in economics, see Deaton 2010, Deaton and Cartwright 2018).

If we acknowledge this, and that no rule benefits all, rule utilitarianism seems to collapse into moral particularism; where each situation must be evaluated individually. On this view, there are no absolute moral rules that always produce a good outcome. Moral particularism is naturally akin to causal singularism, which says that each causal situation is unique with a unique set of causal factors. We should therefore not expect that the same intervention always gives the same outcome. The same intervention in different contexts could bring about widely different outcomes.

What we could do is to allow a rule saying that we should *tend to* favour interventions that benefit the most. But there should be an additional rule that contextual and theoretical evidence must be considered before making a decision. Some might think that this goes against the original motivation of evidence-based decision making, which was to favour quantitative data over expert judgement and mechanistic knowledge. By adding the second rule, we effectively place the decision with the expert, since she is the one best positioned to make an informed decision about a particular context.

If we want an intervention to be beneficial in the context to which it is applied, it seems that rule utilitarianism must give way to moral particularism, and RCTs must give way to expert judgement. By this we are not saying that the rule is superfluous or unimportant, but that it cannot dictate decisions. Rather, the rule should be informed by judgement at least as much as judgement is informed by the rule. For decisions to be based on the 'best available evidence', 'evidence' must thus include more than what we get from RCTs.

PART VIII

Discovering Causes and Understanding Them

24

Getting Involved

24.1 Hands Off

In Part VIII, we will draw together some of the themes that have emerged in the book on the topic of causal discovery. First, we consider the notion of intervention and how it relates to causation. Does it provide a way of uncovering causation or is intervention already dependent on that very notion? Next, in Chapter 25, we explain how much of scientific development involves discovery of causal powers. New technologies, for instance, are premised mainly on the utilization of previously unknown or unexploited powers of things. Chapter 26 emphasizes the importance of experimental failure as a method—or perhaps we should say an *anti-method*—in science. Progress is not, of course, all about confirmation of theories. Typically, corroboration will tell us less than an unexpected result can. Experimental failure is where novelty is often to be found and, thus, presents some of the greatest opportunities for science to stride forward. We then move on in Chapter 27 to consider the rational stance to take, given that we have a host of scientific methods, all with something to offer, but each with at least some shortcoming. We situate this issue within a particular metaphysical framework. Causation is one single thing that cannot be reduced to anything else. It has no absolutely reliable marker in nature. This justifies a methodological pluralism, as others have noted before us, but it does not justify an ontological pluralism. Causation itself is unitary and primitive. We thus contribute a distinctive new position on this debate, between monism and pluralism. In Chapter 28, we consider an issue that has lurked in the background of many other chapters: the difference, and possible gap, between idealizations and real-world causes, which bears on the practical issue of reproducibility. This part of the book thus ties together a number of related issues, and sets the scene for our final conclusions regarding the norms of causal science.

Our first subject in this part, then, returns us to the norm of objectivity in science. We have already seen in Chapter 11 how this norm is compromised. It seems that there cannot be observation without theory. But there is not theory without observation either; the two have to come together. As Popper (1972: appendix 1) showed, even a decision where to look, or what to look at, reflects our interests. Not every fact about the world concerns us. Not every fact is one we could exploit. Not every fact comes to our attention.

The norm of objectivity for science manifests itself in some of the empiricist thinking about causal discovery. We could say that it is there in the very foundations of empiricism. Locke, who is seen as the founder of British empiricism, thought that we start as blank slates, or 'white paper' (Locke 1690: II, i, 2), upon which new experiences are recorded and ideas formed. This means that we approach the world with no prior experiences, theories, or expectations. Bacon (1620: III) added a normative dimension to this, saying that we should begin scientific enquiry with presuppositionless observations in order to avoid implicit biases, which he called the Four Idols. And in Hume (1739: I, iii and 1740), we find this attitude related directly to causation with a hands-off attitude towards causal discovery. Hume's theory suggests that causal knowledge comes as a result of passive observation of the facts, or 'objects', and then noting the cases of repetition where there is constant conjunction accompanied by temporal priority and contiguity. The role of the researcher would be to truthfully record the occurrences. The empiricist ideal, then, is to remain passive in such observations, taking care not to interfere with the course of events, and active only in drawing up the list of regularities.

The ideal remains in some ways of understanding scientific practice with the professed aim that we should record pure, raw data, looking for patterns and correlations. It is regarded as essential, again in a normative sense, that we resist putting our own expectations and interpretations into the observation, as this will take us away from the objective facts. The experimenter must avoid interfering with the result and has to take measures to ensure that her own influence on outcomes is shielded.

The casting aside of prejudices is a theme not just of empiricism but of modern philosophy generally. It is there in Descartes (1641), at the very start of the era, where the danger of possible influence is considered so strong that it is best to forget everything that you think you know and start from scratch. It was easy for empiricists to use such an attitude to attack metaphysics as speculative and, ultimately, useless. Hence, Hume was looking for a notion of causation that was entirely grounded in what could be observed. No causal connection could be seen between two events. We can only think of the former as a cause of the latter because the first type of event is regularly followed by the second type of event. And if no real power, no necessity, no 'oomph', could be observed between one instance of an object being followed by another, then it could not be observed in many repetitions of the same either. Causation is thus reduced away into something else: but something perfectly open to empirical inspection.

24.2 Science as an Activity

So enshrined is the norm of objectivity that it may be wondered how there is any possible credible alternative. Surely we cannot be recommending the pursuit of science in a subjective manner, in which anything goes? Indeed, we are not quite doing that; but we do wish to emphasize that science is an ineliminably normative and social enterprise, which struggles to maintain its impartiality, neutrality, and autonomy (Lacey 1999) and is best to acknowledge its interests. Indeed, it is a practice that would be impossible but for its interests. This view does not undermine scientific realism but can be seen as a foundation of it. The norms that are adopted are ones that allow science to progress and causal discoveries to be made, whereas those things would be impossible were science genuinely value-free. Such a misplaced ambition would surely fall foul of its own standards, much as the logical positivist attack on metaphysics collapsed when it too found non-empirical assumptions inescapable. The claim that science should be free of norms is itself a normative claim (just as the positivist verification principle was itself unverifiable, Hempel 1950). The acceptance of theories that allow us to manipulate the world to our own ends itself shows a value within science.

When it comes to causal discovery, anyone who thinks it's about simply recording the data or mapping the facts is missing the point that scientific discovery is an active process, and especially so when it comes to causation. Our greatest opportunity for causal knowledge consists in the fact that we are able to interact with the world around us. We are part of it: a single, vast, causal nexus that is able to have effects on us and our recording instruments as well as us intervening and having effects on it. As we already argued in Chapter 2, even observation is a case of such an interaction so already this presupposes the reality of causation. And we are ourselves causal agents, empowered to change the world. It follows that Hume's 'perfect instance' of causation is anything but. He describes observed collisions on the billiard table (Hume 1740: 137) as if he is not a part of the same causal manifold but a disconnected 'Godly' observer. Instead, he could have been getting his hands dirty, becoming involved in those causal interactions. And this is what much of science, indeed, does.

A number of philosophers have argued that causation is best understood in terms of intervention or manipulation (Collingwood 1940, Gasking 1955, von Wright 1971, Woodward 2003). Our view on this claim is nuanced. We maintain that causation itself is basic, primitive, and irreducible. One cannot understand causation conceptually through these other notions because they are themselves causal notions. Clearly, then, we could not reduce the concept of causation to one of intervention as the latter presupposes the former. However, the idea of intervention can nevertheless be helpful when it comes to causal discovery. It can also tell us what the point of causal knowledge is, namely that manipulability is why causal knowledge is of use. Woodward puts the point so:

I suggest . . . that the distinguishing feature of causal explanations . . . is that they are explanations that furnish information that is potentially relevant to manipulation and control: they tell us how, if we were able to change the value of one or more variables, we could change the value of other variables. (Woodward 2003: 6)

And 'It also has the advantage of exhibiting an intuitively appealing underlying rationale or goal for explanation and the discovery of causal relationships: if these are relationships that are potentially exploitable for purposes of manipulation, there is no mystery about why we should care about them' (Woodward 2003: 7).

If there is this tight connection between causation and manipulability, then, it has a dual use. It is something we can exploit once discovered, to get what we want from the world; but it is also a connection that we can exploit in order to make our initial causal discoveries. We can change something to see whether it has any effect on something else. This is a hands-on approach. It is not merely that we record a correlation. The idea is that we have intervened to produce the change. It is of our choosing what to change, and when and how to change it, so that we can see whether a second variable changes with it. And this allows us, for instance, to test a theory by making an intervention on the first variable at a point at which the second variable is unlikely to change otherwise. For example, suppose one wished to test the causal hypothesis that the presence of saltwater precipitated oxidation in iron. One could intervene to introduce salt in solution to samples of iron that were otherwise showing no sign of rusting and then wait for the results. One could do this also in a comparative way, contrasting it with iron in a still wet but salt-free environment. The more control one has—control being another implicitly causal term—then the more one is able to gather from positive results.

The example shows how we can use intervention in conjunction with other methods to increase our confidence in a causal hypothesis. That an intervention is followed by a change in another variable counts as corroboration of a hypothesis, using corroboration in Popper's sense of the theory being tested and withstanding that test; that is, not being falsified by the test. However, this can be coupled with a method of comparison: comparing the case of intervention with another in which there was no such intervention. As developed in Chapter 27, we can adopt plural methods to test the hypothesis, where confidence in that hypothesis is increased as the hypothesis withstands tests of more different types. Here we have something like Mill's methodological pluralism (Mill 1843: 454, 455), where he argued that we need both a method of agreement (finding that A is followed by B) and a method of disagreement or difference (finding that B does not occur without A) in order to have confidence that A causes B.

Woodward runs these two methods together somewhat when he explicates his notion of an intervention in difference-making terms. As we said in Chapter 15, we have nothing against the idea of causal connections supporting counterfactual inferences, of a certain type, but we do not agree that causal connections are nothing more than counterfactual dependences among types of event. Were causes and counterfactuals to accompany each other always, a Euthyphro question could still be asked. Is there a causal connection because there is a true counterfactual or is there a true counterfactual because there is a causal connection? We would certainly opt for the latter. However, as we said earlier, we do not think that these two must always accompany each other. Causes tend to make a difference but there can be causation without difference-making (Section 15.3) and difference-making without causation (Section 15.4). This tendential connection is still enough for agreement with Woodward when he says:

an explanation ought to be such that it can be used to answer what I call a what-if-things-had-been-different question: the explanation must enable us to see what sort of difference it would have made for the explanandum if the factors cited in the explanans had been different in various possible ways.   (Woodward 2003: 11)

In our case, we interpret this view as a dispositional one: interventions tend to make a difference. We take this as adequate grounds for answering the question of what if things had been different. But it is possible that an intervention makes no difference, where an effect is overdetermined, and the same effect would then have happened even without that intervention. The notion of an intervention is not an inherently difference-making notion. Instead, we would wish to explicate it in the irreducibly causal terms of production. An intervention on one variable is *able* to produce a change on a second variable and the purpose of experimentation is to discern whether this is indeed the case for two chosen variables. In many cases that will come down to whether such causal production makes a difference, as revealed through a comparison or method of difference; but in other cases it might not. We then have to resort to other methods, such as mechanistic investigation, in order to increase our confidence in a causal hypothesis.

An interventionist approach sits well, then, with a host of scientific practices in causal discovery. Many, if not all, cases of experimental and practical research will involve intervention and manipulation. In genetic research, for example, the experimenter modifies a sequence of a DNA strand and sees what comes of an organism's development, subsequent to that change. In medicine, physics, chemistry, psychology, and at least some parts of social science, experimentation is also founded on our interventions: on changing the situation and seeing what changes with it. Even economics and philosophy have their own experimental branches now, all of which are dependent upon our ability to intervene. It is possible to extend such a notion from actual interventions, that human beings are able to make, to 'in principle' ideal interventions, which in reality are outside of our control (Woodward 2003: 9, 10). One could perhaps think of what would follow if there were an additional planet between Earth and Venus. It is beyond our ability to insert such a body but we can calculate some of the differences that it would make; for instance, that we might have an additional solar eclipse. This is another case where we think that causal powers could explain the difference-making intuitions that we have, rather than vice versa.

24.3 Doing Science from the Inside

We cannot conduct science from the outside, as if we are not a part of the natural world that we investigate. But we should not want to do so in any case. We decide what data we want. We intervene to get it, we gather it, we organize it, we develop methods for analysing it. We choose the focus—the research questions—in accordance with our interests, and we change the world when we start to take a scientific interest in it. This principle is clear in social studies but even in hard sciences, as evinced in Bohr's complementarity principle, stating that the results we get from a measurement of a quantum phenomenon is also determined by the measuring device. The experimental set-up is thus part of the cause in a quantum mechanical experiment, meaning that the result is partly produced by the measuring itself (Bohr 1937: 87). An observation is thus an interaction and the data are already processed by us before we can record them.

What does that mean for the objectivity of science? Is there any way of avoiding a slide into anti-realism? We think so. It helps, for instance, if we are aware of what we bring to the observation so that we are in a position to explicitly and critically discuss the presuppositions at work. No experiment has ever been made that is free of presuppositions so our best practice is to acknowledge what they are. Hence, it is right that the presentation of any experimental result begins with an outline of the methods used, including any philosophical or metaphysical assumptions. The latter might sound outside the scope of science but, as we have argued, philosophical assumptions are always to be found in science. An experiment will also be designed with an expected outcome. A causal hypothesis is likely already to be in mind, which the experiment is seeking to corroborate or, sometimes, to rule out. The experiment in question is chosen precisely because it will have been considered in advance to be a good test of such a theory; for instance, as an intervention on A, looking to see whether B. Here B would be considered otherwise unlikely but for an effect of A, as in abductive reasoning.

It may, then, be impossible to step outside of science to assess it objectively. According to Nagel (1986), even the desire to do so suggests an inherent tension. Science seeks a view of the world but one as if from nowhere, which is impossible since any point of view is a view from somewhere. But an old metaphor can be helpful here. Otto Neurath (see Quine 1960: 3, 4) likened progress in scientific theories to staying afloat in a boat while repairing the planks. Our science rests on the understanding we currently have. Methods, for instance, are premised on how we believe that the world works, such as the view that experience is caused by perceptible objects that can also exist unperceived. Some parts of science may be self-validating, to an extent. But we are aware that some parts of science are incomplete and need additional theories. We can also be aware of weaknesses within existing theories. We have, however, no choice other than to continue staying afloat on existing theories while we try to build something better. An example would be the theory of light, which is deficient in a number of respects. To account for some phenomena, we are best treating it as if it travels in waves. For other phenomena, it makes sense to understand it as travelling as particles. It seems it cannot be both but we can switch between theories as needed.

We can get by on existing theories even though we realize that they could be better or that they are still being improved. Of course, there is a desire to understand, where this knowledge is for its own sake, and for which we want to know the truth of the matter. A key point, though, is that truth is not the only purpose of science. However deficient we might suspect our boat to be, it has to be good enough to keep us afloat. In this case, that means that our science has to be good enough to get us what we want, or as much of what we want as possible: desirables such as food, energy, health, and life-improving technologies. A theory can have some instrumental value, with respect to such goals, even if it is possible to improve it and get even more. As the practitioners of science—the observers and interveners—we bring our own network of prior beliefs, experiences, and interests to the investigation. Contrary to anti-realism, however, we cannot force any view upon the world. At some point it will meet resistance. The theory that the Earth is flat, for instance, faces a never-ending series of anomalous results that require elaborate, ad hoc moves to accommodate. And such a theory gains us no advantages over a round-Earth theory.

We are causal agents and patients. Observations and theories are a mutual manifestation between the world and ourselves. Not any interpretation can be forced upon the world, contrary to subjectivism. But every interpretation includes us as interpreters, contrary to objectivism. Dispositionalism recommends something in-between: depicting a world in which we are immersed as part of the very object of study. New discoveries can still surprise us, however, forcing their way into our attention, defeating our expectations, and creating the possibility of new avenues of enquiry. This is what makes science an ongoing, dynamic process, where more interaction leads to more knowledge. The first step in understanding that process, we have argued, is rejection of those views, especially empiricist in origin, that suggest we are somehow causally disconnected from the world we are studying. This would effectively liken us to supernatural beings, perhaps like Laplace's demon, the super scientist. Instead, we recommend an account in which experience and theory develop in unison, as a mutual manifestation between enquirer and world, which qualifies as a version of realism. Our interactions with the world, as embodied, intelligent, thinking creatures, give us the opportunities to grasp the nature of reality, especially its causal aspects.

25

Uncovering Causal Powers

25.1 Science Changes the World

To think that the role of science concerns only discovery, explanation, and prediction misses something else that is important to it. In Chapter 24, we talked about how discovery is assisted by taking a hands-on approach, causally interacting with the object of study. But there is another sense in which science should be hands on, namely when it comes to putting our discoveries to use. This chapter will be concerned with technology and presents an account of how it should be understood. A distinction can be drawn between science and technology. Skolimowski (1966) suggested that science is about discovering what there is, whereas technology is about what there will come to be or what there ought to be (see also Simon 1969). Similarly, Bunge (1966) pointed out that technology is about action, about changing the world, but it is also action based on theory: the theory developed by science.

We will present an account that supports this sort of distinction. However, while making sense of the above comments of Skolimowski and Bunge, we also see that science and technology can be understood as more continuous than sharply divided. The sharp divide suggests that science is a disinterested pursuit of knowledge for its own sake while technology is about satisfying our interests and getting what we want. While it might be mainly a task for sociology of science, this strict distinction can be challenged. As we have argued earlier, decisions about what basic scientific research to conduct might already be responsive to our needs and reflect the dominant sociopolitical interests. Science is always historically situated. This does not automatically lead to anti-realism, though. There are many truths about the world and we do not have the resources to scientifically uncover them all. The matter of which ones do get uncovered is where economic and political interests can have a role.

Those matters acknowledged, we now wish to focus primarily on the theory of technology. Taking the lead from those earlier theories, we see that technology is about changing the world, rather than merely understanding and recording it. This changing, of course, counts as an intervention, hence it assumes the reality of causation. But it is not an intervention for experimental purposes. Technology, instead, uncovers and then utilizes the known causal powers of things, making some aspect of the world how we want it to be. There is a long history of this: in agriculture, for instance, to increase crop yields; in engineering, to manufacture labour-saving machines; in communications, to give us the internet and smartphones, and in medicine, to create vaccines for disease. We need not labour the point. It is clear that technologies have had a marked impact on civilization, making it easier for some of us to cover our basic needs for food and safety and, going beyond that, enabling us to focus on goals beyond day-to-day survival.

Causation, then, clearly plays the major role in grounding the possibility of technology. Once we understand the causal powers of things, we can harness them in the development of new technologies. And, in other cases, the explanation may go the other way; that is, we could have an idea of what we want to do, technologically, and then try to figure out what we will need to assemble in order to be able to do that. After all, it is not as if we can chance upon some sophisticated machine and then put it to use. The machine will have been built precisely because we had a use to which it could be put. Some cases are of the other kind, however. There are natural resources, such as wind, the sun, and waterfalls, that we exploit because they exhibit causal powers that we realized can be useful. More often, though, there is an intermediate stage in which we realize that a natural resource will be able to do something if we apply some further process that is able to release a useful causal power, which might otherwise remain 'locked' within the resource.

Before saying too much more about these different kinds of case, however, it is probably better now that we give a general theory about what unites them so that we can then apply it back to its real instances. Our theory is one that puts the notions of causal power, function, and possibility at the centre.

25.2 Powers as Grounds of Real Possibilities

There is no denying that discovery is important to us because it satisfies our sense of wonder and curiosity. Human beings, naturally, want to understand (Aristotle, *Metaphysics* 980a). However, that which is discovered is important also because it opens up new possibilities. In some instances, these possibilities will give us some of the things we want. In others, they will open up possibilities that we did not even imagine. Few people were sat yearning for the World Wide Web, for instance, prior to its creation. It took a visionary to imagine it. But once created, and those possibilities comprehended by the rest of us, many took up the opportunities afforded.

But what is the ground of such possibility and how does it relate to causation? Our preference is to explain natural possibilities in terms of causal powers. There is much literature on the metaphysics of causal powers (e.g. Harré and Madden 1975, Mumford 1998, Marmodoro 2010, Ellis 2001, Molnar 2003, Groff 2013, Groff and Greco 2013, Jacobs 2017), but those are largely debates into which we need not enter. It doesn't matter here, for instance, whether causal powers are irreducible or can be explained in other terms. What matters for our purposes is the idea that a causal power grounds a real-world possibility.

Possibility is also a major subject in its own right. There are weak theories of possibility such as the logical theory in which something, P, is possible if and only if it does not involve a contradiction. Hence, on this account, it is possible that boiled water explodes: because it doing so would violate no rule of logic. This theory is of little use for technology, however, which needs something more than the mere logical possibility of its aims. We need also a possibility that is within our power to create. Lewis (1986a) proposed a theory of possibility in which P is possible if and only if it is the case in at least one world, where there is a plurality of other worlds, at which different things are true. There will be a world, for instance, in which salt is not soluble but has different powers instead. Again, this theory is little use in explaining technology to us. The different worlds, in Lewis' theory, are spatiotemporally disconnected from each other. What happens in one world cannot affect what happens in another. Hence, the different powers that salt allegedly has in another world have no bearing on what we are able to do with salt in this world.

These are some of the reasons we prefer a dispositional theory of possibility, which is another name for the theory of possibility based on causal powers. This theory is entirely this-worldly, as opposed to Lewis', and it is naturalistic, as opposed to basing possibility on mere lack of logical contradiction. Causal powers offer us the grounds for what is naturally possible. It may be possible to extend this into a theory of logical possibility but that will not concern us here. The basic idea of the dispositional theory (as, for example, in Mumford and Anjum 2011: ch. 8, and Vetter 2015) is that P is naturally possible if and only if there exists a causal power for it. Hence, it is possible that something burns because there exist a number of different things that have the power to burn. And it is possible that a particular tree burns because it is one of the things with the causal power to burn.

There is an obvious objection to this account of possibility, which comes from a source we have encountered already. It seems that an object with a particular causal power could be in a situation where it is incapable of manifesting its power (Martin 1994). For instance, there could be a bank security system that so far has kicked in every time there was a break-in attempt. In that case, it seems that there could be a power of a thing, to be broken into, for instance, but it is not possible to break into that thing. In line with our earlier account (Chapter 8), we would say that this is because of the presence of an additive interferer, such as the security mechanism. However, our commitment to the dispositional modality (Chapter 10) shows that P is a real possibility even in such circumstances. The additive interferer itself, on our view, will only tend, or dispose, toward blocking the possibility of P. It could therefore fail. And that means that P is still possible as long as the power toward P exists. The objection that P can be prevented from ever being manifested is thus not a serious one for our position. But it would be for someone who thought powers necessitated their manifestations when stimulated (such as Vetter 2015), because the additive interferer would always do its job of preventing P.

This helps convey the general point of our account. There exists a power to burn, for example in wood, hence it is possible that something burns. It may be impossible for some other things to burn, such as water, because they have no power to do so. The existence of powers thus tells us what is and what is not naturally possible. No human has the causal power to jump to the moon, hence it is not possible that they do so, but they do have the power to calculate the distance to the moon, which they have done. Some possibilities have thus also become actualities but not all of them have. It is possible that there be a 100 metre-wide gold sphere but there has never been one. It has neither naturally occurred nor been manufactured. We could build one if we had the desire and resolve but there seems no point in doing so.

However, finding out what are the real possibilities is itself no easy matter. That something is simply conceivable seems to be no reliable guide, for instance, partly because we do not know the constraints on what is conceivable. Some might claim that they can imagine a positively charged electron or a time travel machine. Even if they did, it would not show that any of these are real possibilities. It would only show that impossible things can be imagined. But it is in any case debatable whether a positively charged electron really is imagined. What is such a person actually thinking about? Is she just thinking of a positron and mistakenly assuming it is an electron? If so, that might not be enough to say that she has imagined a positively charged electron. And when it comes to real natural possibilities, the same constraints seem to apply. Consider again whether there really could be a 100 metre-wide gold sphere. Would it fall apart and collapse under its own weight? The answer depends, as we say, on the causal powers of things, which are open to theoretical and empirical investigation. Does that much gold have the power to stay together in a single object?

If powers ground natural possibilities, then how does this relate to technology? The simple answer is that technologies exploit these natural possibilities by using such causal powers or combinations of them to instantiate some desired function. Consider a can opener's design, for instance. There are a number of mechanisms at work in a standard design that instantiate a host of causal powers. The opener has handles so that it can be held by a user and which can come together so that a cutting mechanism can grip the rim around the end of a can. When they do so, a sharp wheel is able to cut, under pressure, into the metal of the can. But, in addition, there is also a cog mechanism attached to a further handle that can be turned. When this happens, the opener has the power to move around the full circumferences of the can's rim, pressing in the sharp wheel as it moves, thus cutting off the whole of the can lid. Needless to say, this combination of powers is not naturally occurring. The can opener is an artefact designed with a specific purpose in mind and, thus, empowered with a complex function. Adding to the complexity, this is a case of what we call complementary technology, which is to say that there is only a need for can openers because there are tin cans that package food. There would be no need for can openers without cans and no need for cans without can openers. The two come together.

25.3 Creating New Possibilities

We can take it as a given that technologies advance over time. New innovations can build on what already exists and add to it. How can this be? In our account, innovations can be of a number of kinds but we could sum them all up as cases of uncovering new causal powers.

The basic kind of case is where a technological advance is a result of uncovering a previously unknown causal power of an already known thing or substance. There are a number of natural resources around us for which we may not yet know the full range of their abilities. There is a good explanation for how this can be so. Some causal powers are easily manifested and can come to our attention without much effort. Water is heavy, for instance, so can be used to power waterwheels. It was not hard to understand this potential. But other powers need more teasing out. The circumstances in which they will manifest themselves may be very artificial and need much work to set up. Martin (2008: 48) says of powers that they manifest when they meet their mutual manifestation partners. In the past, this idea was usually articulated in terms of powers needing to be triggered (see Bird 2007, for example) but we prefer Martin's conceptualization. It suggests that two or more different powers are able to do something together that neither could have done alone. Mutual manifestation also suggests that all the powers are active and more or less equal partners in what they produce. The causal powers within oil provide one of the best examples of this kind, in that the technological breakthroughs did not come from the mere discovery of oil but, rather, in finding ways to bring out and harness the powers within it. The realist about causal powers maintains that all the possibilities were there; but the challenge for us was to find the ways in which they could be brought out. Mainly, this has involved various processes of refinement, to make petrol and plastics, for instance; and then finding the complementary technologies that use them, such as the combustion engine that runs on petrol. Hence, it is upon the discovery of these processes, which transform the original substance, and then finding a complementary technology, that new possibilities become practically available to us. Some technology is more concerned with how to manage all the negative effects of these innovations; the petroleum contaminants, CO_2 emissions, oil spills, plastic garbage, and other negative impacts on the environment, which turned out to be further causal powers of petroleum.

It is clear, then, that there will be different types of processes for uncovering causal powers. A new kind of substance or thing could be discovered that has a useful causal power, such as when diamonds were discovered that had a particular strong ability to cut other surfaces on account of their hardness. No further refinement is here needed to bring out the causal power. The example of oil is slightly different, since it took more technology to bring out the causal powers. A third case is where something completely artificial is synthesized, usually because it has, or is suspected to have, a useful causal power. This is the typical motivation behind the science of pharmacy, such as with 'orphan drugs' that are designed for the treatment of a specific disease. However, this doesn't rule out the possibility that a trial drug, upon testing for one kind of causal power, is discovered to have another one that is even more useful. This happened in the case of Viagra, which was intended to treat chest pain but was found to have another ability as well, and Penicillamine, which was developed for Wilson's disease but found to be effective against arthritis. It is possible that useful mutual manifestation partnerships are discovered by accident, rather than always being designed.

25.4 Functions and Artefacts

Some of these cases can be understood better if we consider the relationship between functions and artefacts. Kroes and Meijers (2006) claim that artefacts have a dual nature. An artefact is a physical object but it is more than that. It has a function, and this shows that it has a design, a purpose, and thus human intentions behind it. The can opener is an obvious example. There are other objects, of course, that were not designed by us, such as rocks and trees, which have a physical nature but no clear purposive nature.

There are a number of ways in which one could criticize this kind of distinction. One could argue, for instance, that natural objects (non-artefacts) can also have a function, the proper function of the heart being to pump blood, for example (Wright 1973). An artefact need not be a physical object, such as if one said that a socio-economic system was a human creation designed with a particular purpose. And a type of substance, such as a synthesized drug, should count as an artefact without being specifically an object. There is a further possibility of an artefact, such as a drug, being designed for one purpose but proving more useful for another, as we have just seen. Perhaps these points can be seen as amendments to a basic theory that technology involves utilization of causal powers of things that would not manifest themselves, in the way concerned, purely naturally. This will make all technologies artefactual in that they would not be as they are but for our interventions. Hence, a process could also be considered an artefact if it makes something able to manifest in a way it would not do naturally (see Lie 2016 for a detailed discussion of naturalness).

There is a question this raises of how we separate something's function from the others of its causal powers. A can opener could also act as a paperweight or be used as a weapon if thrown. But neither of these are its function, just as the heart makes a sound as it beats but its function is to pump. It is for this reason that the account of artefact must contain essential reference to some purpose for which it was designed, and thus why agency must be a part of its account of function. In the natural case, some other explanation will have to be offered (the heart was naturally selected, within some species of organism, for its power to pump, not for its power to make a noise). The case of drugs designed for one purpose, but better suiting another, shows that technology does not always progress according to a rational algorithm, however.

Developments can be unplanned. What remains true is that even if Viagra was originally designed for one purpose, it is now manufactured for a different one. Hence, purpose is still an essential part of the story of the existence of Viagra, even if that purpose is a different one from that of the planning stage.

The variety of examples discussed for uncovering causal powers also shows us that there can be instances related to both things and functions. There are substances or objects that are known to exist, which are then found to have useful causal powers, perhaps after discovering a suitable mutual manifestation partner that brings it out. In other cases, the function is conceived first and only then do we seek to build or synthesize something that fulfils it. In other words, sometimes the thing comes first and sometimes the function comes first. Coal, oil, and wood were all found naturally and were discovered to have the useful causal power to burn. The motor car, in contrast, was built because there was a desired function that was conceived and recognized to be lacking. The technological challenge was to find the type of object that could adequately fulfil that function in the best way.

The issue of epistemic humility impinges on both types of case, regarding thing and function. Given the realism about causal powers that grounds this theory of technology, it is perfectly possible—indeed, a near certainty—that there are objects and substances already available to us that have causal powers we have not yet uncovered or harnessed. We have not yet found the mutual manifestation partners that will release these causal powers. The work to find a cure for AIDS, for example, is premised on the hope that there is something that has the requisite causal power, be it a drug or other kind of intervention. There is, of course, no way of knowing in advance whether there is or isn't such a thing. Second, epistemic humility is relevant where we have not yet had the imagination to think of some function that, were we to do so, might be both beneficial and relatively easy to physically embody in a manufactured object. It was not so hard to build a folding bicycle, for instance, but someone first had to imagine it and decide that it would be useful. The unavailability of solutions to difficult problems could involve a combination of both these types of ignorance.

Given that we have conceded that artefacts, in the broad sense of the term, can have multiple causal powers, only some of which are the reasons those artefacts are brought into being, we see how the possibilities grounded by causal powers include the possibilities of side effects. Ontologically speaking, there are no side effects: only effects.[1] But we can define a side effect as a result of a causal power, usually an undesirable one, of something that exists for one of its other causal powers. It is possible that a side effect can be useful, as in the case of Viagra, but the term is typically used in a negative way. And this, of course, raises issues of risk and its assessment (Rocca and Anjum forthcoming). There is a clear danger that in

---

[1] We got this point from Anna Luise Kirkengen.

designing a technology for one kind of effect, and then testing to see whether it is really there, one ignores completely another effect of that same technology. The motor car is very successful as a method of rapid transport on open roads but, of course, would not be a success overall if it has other very negative effects concerning something else.

These considerations concerning technology clearly have causation at their heart. They show not just that causation is vital to technology as well as science, but that the identification of the causal powers of things remains one of the most important tasks. Science is often justified because it works. It allows us to manipulate the world to our own ends. To an extent, that is true, even though we are capable of being mistaken over what is good for us.

26

Learning from Causal Failure

26.1 The Diminishing Returns of Confirmation

Discovering causes is a hands-on activity, we argued. To see the full causal potential of something requires that we engage in dynamic processes of exploration, continuously seeking to develop the theory in all its complexity and detail. In this chapter we explain how such deep theoretical knowledge cannot progress simply by accumulating positive test results for our hypotheses. If the main focus is on collecting data that confirm the theory, we get a very poor form of causal knowledge, at best identifying A as a cause of B. Should we need to understand *how* or *why* A causes B, however, we cannot rest content with positive results only. Instead, we must also recognize the rich potential to expand our knowledge that lies in cases of causal failure.

When attempting to establish a causal relationship between A and B, we saw that Hume thought the strongest proof of this would be to have our causal expectations constantly and repeatedly confirmed (see Section 5.1). This idea affects the way we usually go about testing our causal hypotheses, through repetition of positive results. A single experiment would thus not count for much, scientifically. We need at least to repeat the experiment a few times before drawing causal conclusions. The same holds for larger studies. One individual randomized controlled trial (RCT) might not count as conclusive evidence, but metastudies that take into account all the relevant studies performed can carry stronger epistemic force. Systematic reviews of existing research are useful in this respect, since they allow us to evaluate whether there is enough positive evidence to confirm a causal theory.

A risk with this approach is that we might become victims of confirmation bias; the disposition to find confirmation of a theory more than is rationally warranted. The explanation for this is partly psychological but can also be found in sociology of science where there are financial and professional incentives for a theory to be confirmed. This type of confirmation bias is well known and already widely discussed by scientists and philosophers (for a recent discussion, see 'Let's think about cognitive bias' 2015, MacCoun and Perlmutter 2015, Nuzzo 2015) and won't be the focus here. But there is a metaphilosophical, methodological sense in which there can be a confirmation bias and that deserves more attention. This is the positivist bias to adopt a scientific methodology that is based on verification through repeated confirmation, rather than other types of systematic approaches. But how much do we really learn about causation from seeking more and more confirmation of a theory? Not that much, it would seem.

If we only got positive results when we test a theory, each repetition would become less and less informative, since the theory would be in want of less and less corroboration. Each new confirming instance beyond a certain point thus has a diminishing return. While further confirmation takes us closer to certainty, the degree to which it moves closer to certainty is less than the preceding case.

We saw in Chapter 19 that the degree of certainty is often measured on a bounded scale between 0 and 1, whereas the amount of corroboration is technically unlimited. We can acquire as much evidence as we like but our theory will never attain maximum certainty. This is why we said that the relation between degree of power and probability was asymptotic. Increased degree of power takes us closer to probability 1, but we never fully reach it. Hence, there is a diminishing return of probability to increase of power. Belief is capable of overdisposing, we said, just like causal powers generally. We now see that evidence can overdispose too. It can get to the point where it becomes epistemically redundant.

If we keep accumulating confirming evidence, then beyond a certain point, each new addition to the confirming evidence adds less to our confidence in a theory than the previous piece of new evidence did (see Singh Chawla 2016). Popper (1959: ch. 10) thought that this is a pointless approach if we want to make scientific progress. But he also thought that final proof of a theory is impossible and that the best we can get is corroboration; that the theory has survived rigorous testing (Popper 1959: 248). In contrast, negative evidence could then make a huge subtractive difference to our confidence in a theory. Popper argued that science advances through falsification of theories, exploiting the alleged asymmetry that, while we can never know a theory to be true, we can know it to be false. This is not the point we want to make here. Instead, we will be arguing that there are ways in which causal failure can be used to advance knowledge other than by falsifying the theory.

26.2 Failed Experiments and Negative Results

Scientific progress seems more likely to happen with surprising, or unpredicted results. Acknowledging the complexity of causation, described in Chapter 7, we should expect that there is much more to the causal story of B than the fact that it was preceded by A. Repeated positive results tell us nothing further about this causal complexity, or of causal mechanisms. In contrast, when negative outcomes disappoint our expectations they might instead direct us towards other causally relevant factors than A and B. How so?

When causation fails, we can investigate what happened. Failure shows us that we need new developments of the theory. It tells us that there is further investigation possible into the reasons for the failure. The case thus represents an opportunity for developing our causal knowledge in more detail. This point is also recognized by

Cartwright and Hardie (2012) when they discuss what kind of causal knowledge is needed for making policy decisions. Such decisions usually require that the results of a study from one context can be plausibly expected to apply elsewhere, in another context. RCTs cannot provide such a guarantee. When counting the proportion of cases in which A was followed by B, and comparing this to how often B occurred without A, the attention is on the positive results, not the negative ones. But if we look closer into why A failed to give B, we might learn more about the causal mechanisms. Once we understand causal mechanisms, they argue, we are in a better position to judge whether the results from one context could also be applicable here.

The difference between our approach and the RCT approach is clarified by seeing what you get out of failure. In the case of just an RCT, a pure RCT, nothing, because you are not thinking about how or why, just whether, it works. But for us, failure makes us look harder and see why. It helps us.   (Cartwright and Hardie 2012: 130, 131)

As a method for theory developing, paying attention to causal failure is not new. Scientists already use such results to advance causal knowledge. Dunbar (2000) offers descriptive rather than normative evidence, but with some normative implications. He reports that unexpected findings are a main concern for experienced scientists when building causal models. After following the works of scientists in eight different immunology and molecular biology labs, he found that most of the time spent on causal reasoning was used discussing those experiments that failed to give the predicted result. Yet, he notes, the epistemic value of unexpected findings is not an explicit part of science education, meaning that most students do not learn how to deal with them (Dunbar 2000: 53). One of the conclusions he draws from his study is that, when educating science students, one should also teach them to follow up unexpected results, especially in the control conditions.

So how exactly can an experiment that fails to produce the predicted outcome offer new causal knowledge? Rocca (2017) discusses one example in which she notes that the understanding of causation progressed as a direct result of causal failure. The case is taken from a study performed by Ordinario and colleagues on the causal role of the SATB1 gene for the production of breast tumour (Ordinario et al. 2012). The gene was known to cause tumour when overexpressed in a certain type of cell, by radically reprogramming the gene expression and driving metastasis. Yet, the details of the underlying causal mechanism were not known. In this study, cells from two different sources were used and, to the researchers' surprise, the cells from one of the producers failed to give the expected results each time. The cells were genetically indistinguishable, so a difference in outcome was unexpected. A further gene analysis revealed that the cells had been cultured in different mediums, leading to a difference in amount of ATM protein in the cells. New experiments were performed and the team discovered that ATM counteracts or suppresses the activity of the SATB1 gene, preventing the formation of the tumour. In this case, Rocca notes, what led to new causal discovery was not the repetition of the positive result, but further analysis into the context that produced the negative result (for more examples of causal insights from failure, see Rocca et al. forthcoming).

This also shows why Popper was wrong to say that a single counterexample falsifies a theory, at least in cases of causation. Instead, much can be learned from negative results, if we are willing to investigate them further. Recall that the Quine-Duhem thesis suggests that a negative result can never definitely falsify a theory but instead tells us that there is something to be explained: some factor not taken into account (Quine 1951, Duhem 1954). We can now see how this could be used to advance causal knowledge. To find elements that can interfere with the causal process is essential for theory development. One thing is to understand which factors contribute causally to an outcome. This is valuable in itself for the understanding of that phenomenon. But it seems equally valuable to find out what could interfere with or prevent the outcome. Contributors and preventers are thus two sides of the same causal story, and both help us reveal the causal nexus of relevant factors and their interactions.

In assessing risk in particular, understanding causal preventers and interferers is crucial. If a new drug is put on the market only because it is repeatedly confirmed to produce the predicted effect, we do not know the full causal story of how it does so. What we do know, is that the drug does nothing in isolation, since it can only produce its effect through causal interaction with the biochemical processes in the human body. These mechanisms are important to understand for purposes of risk assessment in pharmacology. We thus need a different approach for evaluating risk than when we evaluate a drug for effectiveness (Osimani 2013, Landes et al. 2018). When we learn about some unpredicted effect of the drug, we also learn more about the causal mechanisms.

We saw in Chapter 25 that an intervention can end up doing more harm than good in certain contexts. To uncover the potential for harm is thus equally important as finding potential benefits. For this purpose, it is not sufficient to test only the positive effects of the intervention. We need negative cases, or foils, in addition to the positive results, since these can direct our further investigations and point us to yet undiscovered causes and causal mechanisms.

26.3 Non-Monotonic Causal Reasoning

We argued in Part III that causation is essentially context-dependent. Different contexts involve different causal factors, some of which might counteract the effect. To constantly confirm our causal hypothesis might be useful for identifying a causal factor of an event, but it will not bring us closer to a causal understanding and explanation.

Take a simple example. We know that flicking the light switch causes the lamp to light. This can be tested by repeating the flicking of the switch until we are confident that the lamp always lights as a result. But if this is all the information we have, we cannot say that we understand how the light switch causes the lamp to light. In contrast, if the lamp fails to light when we flick the switch, we can investigate the causal context. Perhaps we learn that the bulb is burnt out, that the plug had fallen out of its socket, that a fuse had blown, or that the wire had been cut off. Each of these failures reveals something about the workings of the lamp. In contrast, if there were nothing we could do to intervene or interfere with the effect, we would learn very little, if anything at all, about causation (see Chapters 24 and 25). If the lamp still lit after we had broken the bulb, switched off the fuse, pulled the plug out of the socket, and cut off the wire, the otherwise confirmed causal hypothesis might have to be discarded. In fact, we might be sceptical of whether we are dealing with a case of causation at all, or rather with some unusual case of non-causal necessity or a prank.

Recognizing that causal processes are sensitive to context should make us more interested in uncovering and understanding the local context in which causation happens. Without such understanding, our causal predictions would then also be less reliable. But again, this comes down to the type of causal question we are asking. A difference can be drawn between 'what' and 'how' questions about causes. The first is about *whether* A causes B, while the other is about *how* A causes B. In a study performed by Ahn et al. (1995), it was found that science students emphasized different types of evidence for the two types of question. When asked to identify a cause, they would look to statistical evidence about covariance between A and B. But if they were asked to explain how B happened, they typically preferred mechanistic evidence. This difference also reflects the way in which quantitative and qualitative studies tend to deal differently with cases of causal failure. In statistical methods, failure is typically dealt with as exceptions, noise, outliers, or simply a negative result (see Section 5.3), while experimental lab settings and single case studies are better suited for investigating why a failure occurred.

In Chapter 22 we saw that the context-sensitivity of causation makes our predictions inherently fallible. But the more information we have about the causal context and the causal mechanisms, the better equipped we are to make good predictions. Still, predictions remain defeasible, even when the causal mechanisms are known. Additional and unaccounted-for factors could defeat them. Any open system will be vulnerable to such interferers, even if they are excluded from the causal model. Hence, it might be that A causes B. But this does not mean that A together with C also causes B, as we saw above in the case of the SATB1 gene in combination with the ATM proteins. While the tumour was caused by the gene, it was counteracted by the presence of the ATM protein. The original causal model only included the SATB1 gene in isolation from the ATM protein, but with the new information, the model was revised.

Causation thus requires non-monotonic reasoning. This is a dynamic form of reasoning, needed to deal with incomplete, uncertain, or changing information. Such reasoning is typically required in diagnosis, law, artificial intelligence, and interpretation of meaning (Antoniou and Williams 1997). Since most of our causal knowledge will be incomplete or changing in this sense, we should always be prepared to change our causal conclusions and predictions. We might have to withdraw, revise, or even provide other causal theories when faced with new information. One suggestion to help keep an open mind in the research process is the multiple-theory approach, of considering the truth of many different hypotheses at once, from start to finish. So rather than trying to corroborate a single hypothesis, with the risks of confirmation bias and tunnel vision that entails, this method of entertaining more working hypotheses encourages us to take into account a wider range of evidence while continually refining their theory (Rosen 2016).

In this process, cases of failure are an opportunity to learn more about the causally relevant factors that interfered and interrupted the outcome. Failed prediction should not automatically be taken as a sign that no causation happened at all, although this might be the case. It could instead mean that there were more causes involved than we had taken into account in our model. Even though parts of the workings of a causal element might be identified by investigating it in isolated settings, we also know that what something does in isolation might be very different from what it does in an open system. While causal models are usually about the isolated context, failure typically happens in open systems and because of contextual interferers. Context-sensitivity of causation thus carries with it a great potential for learning. Rather than being treated as an interruptive element to be expunged from our studies, causal interference should be embraced as a valuable source of new knowledge.

26.4 Mind the Gap

There is a gap between what we already know, or think we know, and what we might learn from new experience. This gap usually becomes clear when something unexpected happens that we cannot explain. When our causal model is challenged by causal failure, we have a case of what Lindseth (2012) calls a discrepancy experience. It is a surprise experience that the world does not fit our expectations. This leads to a crisis, but it is not something to be avoided. Such discrepancy experiences are a condition for learning, he argues, and if we try to avoid being challenged, we also prevent ourselves from learning something new. Instead we should welcome discrepancy experiences, since they make us wiser. They help us develop our knowledge, judgement, and skills. By being challenged in our pre-existing beliefs, our ability to reflect critically is enhanced. Since we struggle to avoid discrepancies, we try to dissolve them. In doing this, however, we are forced to investigate closer what we thought we knew until what we could not explain is incorporated into our theory. This is how theories develop, and it is how we make sense of the world again.

Formulating, revising, and developing causal theories is thus analogous to a hermeneutic process. Typical for these processes is that parts and wholes must be understood and interpreted in light of each other. Sometimes we zoom in on a single element and try to isolate it. Something can clearly be learned from this, such as when we isolate a causal factor to study its causal role in a controlled setting. But we must not forget to also zoom out again and take into account its interactions within the wider context of the causal nexus. Only by considering both parts and wholes together will we be able to expand what Gadamer (1975) calls our horizon of understanding. He thought of understanding as a fluid, ongoing process. Our horizon is constantly challenged and changing in the meeting with the world, and there is no final point of completion where we have full understanding. Learning about causes in all its complexity might also be an open-ended process.

Cases of causal failure could, if approached in the same way, expand our knowledge. If we try to avoid them, or disregard them, we learn nothing new. The history of science shows many examples where scientific discoveries and innovations happened by chance, accident, or luck, sometimes while the researchers were working on something else entirely. When stumbling across a surprising phenomenon, some scientists go on to develop it into a new discovery. The discovery of penicillin was the result of an unwashed Petri dish, allowing the bacteria to grow freely, and the invention of the microwave oven started with the observation of a scientist's chocolate bar melting when passing a magnetron machine. Radioactivity was discovered while working on uranium, sunlight, and photographic plates, after leaving the experiment because of cloudy weather. Sometimes, even failed medical research can turn into a huge commercial success, such as Coca-Cola, which originally was an attempt to find a cure for headaches.

When a scientific discovery is made by accident in this way, we call it serendipity. This concept involves a chance discovery, but also the wisdom and creativity to make good use of it (see Copeland 2017 for a discussion of serendipity). Instead of ignoring the unexpected outcome and moving on with their planned research, these scientists grasped the opportunities that the results presented for new knowledge and alternative applications. In their study, Dunbar and Fugelsang (2005: 712) found that, rather than being victims of the unexpected, the scientists created opportunities for unexpected events to occur and to be followed up with further investigation. By embracing the potential for new knowledge that lies in unexpected outcomes, events to which others might not pay attention could turn out to be serendipitous experiences. If, in contrast, cases of failure are treated as nothing but negative results, not even to be published or investigated further, we might be missing out on a valuable source of information.

Positive evidence is of course valuable insofar as it helps establish causation. In this chapter we have tried to show that cases of failure can be at least equally valuable because they contribute to theory development and new causal hypotheses. If we are taught to repeatedly confirm theories and ignore negative outcomes, we only learn a small part of the causal story. Causal understanding, we have argued, requires that we also have knowledge about causal mechanisms, causal complexity, possible interferers, and the local causal context. So while we might establish causation by way of confirmation, we can understand causation by way of failure.

27

Plural Methods, One Causation

27.1 Imperfect Methods

In the course of preceding chapters, we have come across a number of different approaches to the discovery of causal connections. We have at various stages had critical things to say, indicating that no method is perfect. What, then, should we conclude? That scientific knowledge of causes is impossible? Are we advocating scepticism? We are not, as we explain in this chapter. The shortcomings of some familiar methods, however, do require an interpretation. We need to understand why there are multiple different methods that we use when trying to establish causation. We will develop a justification for doing so and also explain how the use of different methods makes causal knowledge possible, even if they remain fallible. Notably, we will deny another form of scepticism too: scepticism about whether there is a single thing called causation. In response to the use of different methods for discovering causes, and a failure of philosophical analyses of causation, it is becoming more acceptable to settle for a position of causal pluralism: the idea that causation is not one single thing but many. This move can be resisted. There is another alternative, which we will advocate. Causation is, after all, one single thing but, because it is primitive and irreducible, we best approach it via a number of different methods. Hence we should be methodological pluralists, as concerns evidence, while being ontological monists about the nature of causation itself. This distinguishes our position from other pluralisms.

27.2 Causal Pluralism

Causal monism—the idea that causation is one single thing, identical in all its instances—is what the causal pluralist denies. This could be because the pluralist thinks there is no single and satisfactory analysis of the concept of causation. Or, even if there were an adequate conceptual analysis of causation, it may be thought that there was no one single thing in the world that satisfied the concept (see Dowe 2000: ch. 1 for this kind of distinction and Reiss 2015: 203 for its application to types of pluralism). The connection between conceptual and ontological or metaphysical pluralisms is a complex one. A conceptual pluralism would seem to entail an onto-logical one but not vice versa. We will not go into detail on this matter, however, as our task is mainly to support a metaphysical monism despite acceptance of an epistemic pluralism about causation.

Causal pluralists such as Hall (1994), Cartwright (2007a), Psillos (2010), and Illari and Russo (2014: chs 23, 24) deny that any single conceptual or metaphysical account of causation succeeds. Not every cause involves production, they might say. Think of causation by absence (Chapter 18), which seemingly does not. And not every cause involves counterfactual dependence (Chapter 15). Think of overdetermination cases. Now these considerations are well taken. Causation does seem to resist analysis. But how should we respond to that seeming fact?

A possible response is to continue the search for the one true, but so far elusive, reductive analysis of causation. Perhaps there will be an eventual success to one reductive project but we doubt it: not just through pessimism but because we think there is a reason in principle why there will always be a distinction between causation and something else that is not causation (Mumford and Anjum 2013: ch. 8).

The other response, then, is to adopt a pluralist theory of causation. Because the analyses fail, one could say something like the following. Causation is disjunctive: it is one of two things. It is, for instance, difference-making *or* influence. This is close to Hall's (1994) view. Causation would be a disjunction with two disjuncts. But other theories have been open to more than two things counting as different kinds of causes. Psillos (2010) has more. Cartwright (2007a) allows for causation to be lots of different kinds of things and we will discuss her version of pluralism in more detail, mainly because her view is otherwise the closest to our own, but for its pluralism. Cartwright certainly believes in causation, but not that it is one single thing, as is clear in the following statements:

there is no single interesting characterizing feature of causation: hence no off-the-shelf or one-size-fits-all method for finding out about it, no 'gold standard' for judging causal relations.

(Cartwright 2007a: 2)

Under the influence of Hume and Kant we think of causation as a single monolithic concept. But that is a mistake. The problem is not that there are no such things as causal laws. The world is rife with them. The problem is rather that there is no single thing of much detail that they all have in common, something they share that makes them all causal laws.   (Cartwright 2007a: 19)

She then summarizes her conclusions:

1. There is a variety of different kinds of causal laws that operate in a variety of different ways and a variety of different kinds of causal questions that we can ask.

2. Each of these can have its own characteristic markers; but there are no interesting features that they all share in common.   (Cartwright: 2007a: 19)

A possibility that we think Cartwright overlooks, primitivism about causation, is the one we will advocate and, we argue, it is perfectly consistent with the adoption of plural methods of investigation. Before that, however, it might be wondered what is wrong with causal pluralism. Why not accept the pluralist position?

What we take to be the decisive objection is basically a Socratic one. There has to be something in virtue of which all these different things are of the same kind. There has to be a reason why they are all types of causation specifically, rather than types of something else. Now it might be countered that such a question betrays some form of essentialism that the late Wittgenstein (1953) rightly banished from philosophy some time ago. Indeed, Psillos' (2010) answer to our question is that causation is what Wittgensteinians speak of as a family resemblance concept. Different instances of causes are like members of a family. There might not be one feature they all have in common but you can tell they belong to the same family. Some have a similar nose. Others don't, but they have similar hair, and so on. However, this kind of answer is unsatisfactory, as Pompa (1967) subsequently showed. Resemblance alone is not enough because someone might resemble the family without really being a part of it. So it is also crucial that the resembling person actually be part of the family. But this means, for the present case, that it is not enough merely to resemble causation in order to be causation. Some things do that without being causation, such as *sine qua non* necessary conditions (see Section 15.4). The resembling thing also has to be causation, which means that the question of what makes the supposed instances specifically instances of causation has not been answered after all.

There is another kind of answer to what makes all the instances cases of causation and this is the role that they play: the use to which we put them. Williamson (2006: 69) and Reiss (2012) have this view but a similar idea can be found at least as far back as Ayer (1963) (for details of the view, see Illari and Russo 2014: ch. 18). We call something a causal law if it is used as a basis for our inductive inferences and predictions. For example, if we maintain that chocolate kills dogs, we should be prepared to predict that if a dog has just eaten chocolate, then it will die, or there is a significant risk that it will die. For a time, this was also Cartwright's response to the question we are asking. She argued that there are lots of very specific, 'thick' causal claims that we can make; for instance, that someone is *washing* the dishes, that money supply *increases* inflation, that viruses *give* you a cold, and so on. Apart from philosophers, people rarely use the explicit 'thin' concept: that one thing *causes* another. The specific claims are united as 'causes' only because of their epistemic and pragmatic role. As Cartwright says, 'representing a set of relations as causal should allow us to make some kinds of inferences that allow us to use causes as strategies for producing effects' (Cartwright 2007a: 46). However, she goes on to add that this was only her hope and belief in earlier work (Cartwright 1979). Now she is more sceptical that causal knowledge can deliver effective strategies: that it can tell us much about how to use causes. This is because she thinks causal knowledge is produced in very different circumstances—in the laboratory, for instance—from the circumstances in which we have to use it (Cartwright 2007a).

There is a further reason why we should be sceptical about this kind of response to what makes the many different and diverse instances all instances of causation. This is again the Euthyphro question. Are these claims really causal simply because they are useful in explanation, inference, and prediction? Or are they useful for those purposes because they are real cases of causation? We think the latter is the obvious right answer. We use knowledge of a causal connection to make predictions precisely because it is a claim of causation, which, if right, will make such a prediction to some degree reliable. The epistemic solution to the problem of the unity of causation thus puts the cart before the horse. It suggests that some epistemic practices simply are useful, while neglecting the obvious reason why: which is that they are based on causes. This epistemic view of the unity of causation also implies that unless human beings existed, with their epistemic practices, nothing would cause anything else. This view is, therefore, anti-realist about causation itself.

Our conclusion is that there is no satisfactory answer, offered by the pluralist, for why a group of distinct things would all count as instances of causation. Let us, then, consider one other possible response to the failures of analyses of causation.

27.3 Causal Primitivism

The primitivist thinks that while causation cannot be analysed into one thing, it cannot be analysed into a disjunction of several things either. It cannot be reduced down to anything that is non-causal. Causation is real, it exists, and is one single thing that is primitive.

It is not just the failure of proposed analyses hitherto that makes us opt for the primitivist position. Nor is it merely the failure of causal pluralism to explain what unites all the instances as cases of causation. In addition, causation sounds exactly like the sort of thing that would be basic, irreducible, and fundamental. What could be more fundamental than causation? Already we have argued that without causation, there could be no intervention in the world, no use for our knowledge, no coherent order in the world, and, crucially, no observation (Chapter 24). Hence, the project to observe and record correlations is itself premised on the metaphysical belief that causation must be real, otherwise the correlations tell us nothing. Every theory must have some primitives: things that have to be assumed and which cannot be proved on the basis of other truths. What more plausible candidate could there be, for a basic assumption grounding scientific knowledge, than causation? Russell (1948) came to see this. And even in Locke's (1690) empiricism, there was an acknowledgement that while some ideas are built out of others, and thus reducible to them, other ideas had to be primitive. We have argued elsewhere (Mumford and Anjum 2011: ch. 9), so shall not repeat it here, that as agents we can have direct experiential knowledge of causes. We can thus know of the existence of causation, in at least this case, because we act through it and have causes act upon us, again, without it being reducible to something else.

Acceptance of primitivism does not mean the end of all discussion, however. It means that causation cannot be reduced to something else, such as constant conjunction, probability raising, or difference-making. But there still can be useful and informative, non-reductive things that we are able to say about causation, such as that causes tend to make a difference, that they dispose toward their effects without necessitating them, that they are context-sensitive, and so on. We can also say useful things about the epistemology of causation: how we come to know about its instances. This is what we will now go on to do.

27.4 Causal Symptoms

We will spell out a philosophical foundation for a pluralist epistemology of causation. We remain ontological monists about it; causation is one single thing. But we accept that causal knowledge requires plural methods, what Reiss (2015: 203) calls evidential pluralism. It may also be worth noting that Cartwright could adopt this same stance, if she were willing to drop her commitment to ontological pluralism, and it would come at little cost since it is in agreement with much else that she says.

Given what we have argued in this book thus far, we think it right to conclude that nothing else distinct from causation is an entirely reliable mark of it. If there were such a thing—a perfectly reliable indicator—then causal discovery could be a relatively easy challenge. Suppose that there was one thing that always came with an instance of causation and never came with anything that was not causation. We would know whenever we found the perfect indicator that we had found a causal connection. But, as we have shown, there is no such thing. Causes may tend to make a difference, to raise the probability of an effect, to issue in a raised incidence of that effect, to be accompanied by explanatory mechanisms, and so on. But we can have causation without these markers, and sometimes the markers without causation.

Nevertheless, there is a connection that we are able to exploit. We can consider these features not as perfect indicators but as symptomatic of causation. A symptom is a sign that tends to accompany some other phenomenon such that it can be taken as evidence of its presence. Pain in the jaw is a symptom of an abscess, for example. A swollen neck is a symptom of mumps. We will define a symptom, S, for our purposes, as a recognizable phenomenon that tends to accompany some other phenomenon X, so that, although there is a tendency for S and X to come together, it is still possible to have S without X and X without S. Even with such a connection, S can be useful for the identification of cases of X. We deny that there is a strict, deductive inference available from the symptom to the thing of which it is symptomatic. But if the symptom tends to accompany the phenomenon, then a non-deductive inference to it may be warranted. And often, we allege, the position is better than that. In the case of causation, it appears to have plural symptoms, $S_1, S_2, S_3, \ldots S_n$. The more of those symptoms that are present, then the more confident we may be that we have identified a cause. Cartwright has already expressed sympathy for this kind of approach, in one place noting, for instance, that probabilities are symptoms that could be used to identify causes (Cartwright 2007a: 62).

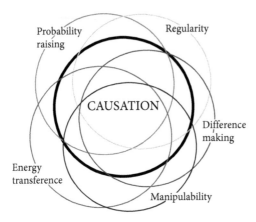

Figure 27.1 Causation and its symptoms

The picture this symptomatic approach suggests is one in which causation is at the centre—the object of our consideration—but because it is primitive and irreducible, it has to be explored via the symptoms surrounding it, where each symptom overlaps roughly but not exactly with the reality of causation. This is depicted in Figure 27.1, which is not intended to show exhaustively every symptom of causation but only to give a visualization of the structure of the proposal. It indicates that causes tend to make a difference, but allows that there could be some causes that don't (overdeterminers) and some difference-makers that are not causes (necessary conditions), and similarly for all of the other symptoms. The symptoms of causation would include the features of context-sensitivity, dispositional modality, complexity, propensity, non-linearity, vulnerability to interference, and so on, which we have added in this book.

27.5 Methodological Pluralism

We talked above about the symptoms of causes rather than methods of their discovery. But we hope it makes clear why there is a route from symptoms to methods. Behind the usefulness, and thus validity, of a method is that it in some sense 'latches on' to one or more of those symptoms. Insofar as randomized controlled trials (RCTs) can tell us something about causation, it is because they adopt a method that exploits causation's difference-making symptom. The reason why comparative methods are useful for this purpose is precisely that causation has this symptom. To give another example, studies using correlation data seem to exploit the regularity symptom of causation, seeking cases where one variable tends to change with another. Again, our view rules that this offers no guarantee of a causal connection, but it could be some evidence of one, especially if other methods give the same verdict. Suppose we do not merely record a correlation but also find that the correlations hold under interventions. Then we may start to gain increasing confidence that we are in the presence of a causal connection.

But how do we know that a method is useful? How do we know that the method has latched on successfully to a symptom of causation? And what sorts of standards do we think should be met by a candidate method before we adopt it? The answers to these questions could be very complicated but we will proceed to offer a general and normative answer for which a number of later caveats and nuances will be possible. In the case of merely identifying that A causes B, as opposed to how or why A causes B (to which we will return later), we should expect of a method that it is able to identify causal connections better than merely guessing, or deciding on the toss of a coin. If a method gets it wrong, whether A causes B, more than it gets it right, for instance, then we shouldn't adopt it. We can think of this as a default principle, of which there could be special cases.

The matter cannot be quite this simple because we want of a method not just that it judges correctly when there is causation, but also that it judges correctly when there isn't. This is because one 'method' that is guaranteed to identify every case when one thing is a cause of another is simply to pronounce that everything is a cause of everything else. The obvious drawback of this approach would be that although it successfully identifies the cases where there is a causal connection, it also misidentifies countless cases as causal that really are not. A good method can identify both because it is just as important to know that two things are not causally connected as it is to know that they are. What we want, then, are methods that can successfully identify both the positive causal facts, and the negative ones, better than pure chance or guesswork. Of course, the abilities of various methods to do this could come in degrees, as we shall consider shortly in relation to evidence hierarchies. A method that was only slightly better than guesswork at identifying the causal facts would rightly still not be considered all that good a method. It also follows from our primitivism, as outlined, that no method can be expected to get it right all of the time.

To give a flavour of the sorts of further nuance we would add, it seems possible, for example, that there is a method that is not all that successful at identifying causal connections overall, in general, but which might be very good at identifying them within some limited sub-domain. Consider energy transference. Some think that it is the mark of causation that it involves the transference of a conserved quantity from cause to effect (see Fair 1979, Salmon 1984, Dowe 2000 and Kistler 2006). We deny that this is all there is to causation but there might nevertheless be some areas in which energy transference is a reliable symptom of it. The question of energy being transferred seems of little importance in the case of a kiss causing someone to blush. But it might be a reliable indicator of causation in the branch of physics known as dynamics, which concerns movements and collisions of objects.

Adoption of plural methods for discovery of causes seems justified, therefore, if those methods are more or less reliable indicators of the presence of causation in virtue of identifying one or more of its symptoms. We thus have a methodological pluralism as opposed to an ontological one. Indeed, our adoption of methodological pluralism is in response to the primitive and irreducible nature of a single, unitary causation, and the epistemic challenge that it raises. We acknowledge that there have been other methodological pluralisms before ours but they have not necessarily been the same, nor with the same motivation, as what we advocate. Ours might be a special case of Feyerabend's (1975) methodological pluralism. His motivation is to tolerate plural approaches so as not to lose out on the possibility of knowledge. This is the same as in Mill's (1859) advocacy of liberty. But we do not allow that 'anything goes'. Some methods might indeed be little or no use in identifying causes. In Longino's (2002) methodological pluralism, any single approach to science will be incomplete, she argues, so there can be more than one correct scientific account of the world. In contrast, we lean more in the direction of realism. Even if there is one correct account of the causes in the world, we still think we need plural approaches in order to identify it empirically.

27.6 Evidence Hierarchies

The best situation we can be in, epistemically speaking, is where we have evidence from a number of different methods that all converge on the same verdict: all deciding that A is a cause of B or that C is not a cause of D. For example, we might have a raised incidence of some phenomenon B whenever A, where the A and B correlate, there is evidence that when A is taken away, B is taken away and, perhaps along with other things, there is mechanistic evidence indicating a plausible causal path from A to B. Even with all this, there can still be no absolute guarantee that truly A causes B. Even the best evidence could be wrong; but it will not tend to be. Causal theories, like any other theories in science, are fallible even when grounded in good evidence. On the other side, we of course allow the possibility of causal connections for which there exists no evidence at all. These sorts of case are possible once we accept realism and primitivism about causation. And what would be the alternative to accepting such consequences? To assume that under certain conditions we can have infallible causal knowledge? That there is an algorithm that gets us from evidence to a guarantee of a causal connection? That is not *prima facie* plausible.

But what of cases when we do not get an agreement from all the different methods? What if one sort of method says that A causes B but another method fails to confirm it? What should we conclude? Or what if only one sort of method has ever been used to test a causal claim so there is only one type of evidence for it? It is because of such questions that we are tempted to form evidence hierarchies, which present us with a rank ordering of the strength of the different types of evidence. The pyramid of evidence-based medicine is one such example. The pyramid tells us that expert opinion is not to be as highly ranked as case studies, for instance; but an RCT is better evidence than a case study, and systematic reviews of RCTs are even better (Howick 2011: 5). Hence, if there is disagreement among methods, or incompleteness of some types of evidence, according to the norm of the pyramid one should always side with the evidence from the highest-ranked category.

We do not want to deny that some forms of evidence can be better than others. We have already accepted that different types of evidence could be better or worse as indicators of a causal connection. But, and without committing to any particular hierarchy, we think there is more than one way in which to understand the weighting of evidence. We do not have a stark choice between a strictly ordered hierarchy and a completely flat picture in which any form of evidence is as good as any other. We could instead have a dispositional hierarchy in which the 'better' evidence tends to outweigh the 'lesser' evidence but not in every instance. The difference between this and a strict hierarchy can be seen in the following type of case. In a typical evidence pyramid, the higher-ranked evidence always trumps the lower-ranked evidence. On our account it is possible that the higher-ranked evidence offers one conclusion but that it is rational to reject it if all the other, lower-ranked evidence is available and contradicts the finding of the one other method. This will depend on exactly what the evidence is and how much of it there is. For instance, there might be only one RCT available and it is set against multiple cohort studies, correlation data, and very detailed mechanistic knowledge. Even if the RCT were ranked higher, it could still be outweighed.

Another concern around evidence hierarchies is that just as we are pluralists about forms of evidence, it seems plausible that we should also be pluralists about hierarchies. The reason for this, as suggested by Osimani (2013), is that how significant the type of evidence is will be in part a function of the question being asked. There is often a focus simply on the question of *whether* A causes B. But when it comes to *why* and *how* questions, the evidence that helps us answer *whether* might be fairly useless. As we said in Chapter 14, mechanistic and qualitative evidence should be best to answer *how* and *why* questions even though it tends to be lowly ranked in a standard evidence pyramid. We should be pluralists about evidence hierarchies if we are being pluralist about the sorts of questions that can be meaningfully asked in the causal sciences.

To the pluralisms already described, it is sensible to add others. We advocate pluralism about methods but also about methodologies. It will be recalled that a methodology concerns overarching principles of investigation: whether one is an inductivist, falsificationist, believes in Lakatosian research programmes, Kuhnian paradigms, heuristics, or whatever. Without saying that each of these approaches is as good as any other, there still could be a case for employing them all if no one of them is capable of revealing the whole truth. Similarly, plural approaches might be warranted in scientific reasoning, by which we mean deduction, induction, abduction, reductio ad absurdum, extrapolation, interpolation, analogical thinking, and so on. Again, none of these forms of reason can be expected to hold a monopoly on truth.

We have presented a thesis that combines epistemic or methodological pluralism with an ontological monism and primitivism about causation. We offer this by way of explanation of our critique of many of the methods of science. None of them offers us a complete picture but neither do they offer us nothing at all. Of course, the claim that a particular method is successful at identifying the causal facts can itself be questioned. How can we know that a method identifies the causal facts correctly when we have nothing but such methods to show us what the causal facts are? This is a puzzle for any realist theory of science, which offers a claim that the theory can be compared favourably to the reality. But we are not alone in saying once more that theories are developed tentatively and bit by bit, building a picture of reality and improving it gradually.

To this, we have added the argument that the best methods will be those that latch on to the proper symptoms of causation, and there has been error about what those are. This perspective provides a new way of looking at the problem of scientific methodology. We are not simply saying that those methods are right that produce results corresponding to the causal facts. The causal facts, whether A causes B and so on, are precisely what we do not know in advance and we adopt scientific methods in order to discover them. We cannot, then, require that the causal facts must be known in advance, as the basis on which the methods are adopted or not. Rather, we are saying that methods should be adopted that correspond to the real, metaphysical nature of causation itself, discoverable through philosophical investigation. If we do that, we maintain, we will have methods that exploit the true symptoms that causation manifests. And it follows that those methods will have a tendency to get the causal facts right more than methods based on a mistaken view of causation and its symptoms.

28

Getting Real about the Ideals of Science

28.1 Reproducibility: a Cornerstone of Science

In this book we have discussed and challenged a range of established norms of science from a philosophical perspective. We have tried to show that reflecting upon how we should understand causation is not simply an intellectual exercise for philosophers, but something that significantly influences how science is shaped and practised. Now we turn to what might be thought of as the cornerstone of scientific method, without which the whole of science could collapse into subjectivity; namely, the reproducibility of research.

Reproducibility can mean slightly different things in different disciplines and for different researchers. But the general idea is that a study is reproducible if it can be carried out in the same way by other researchers and deliver similar results. Unless our studies can be performed by other scientists, and our results can be obtained elsewhere, what are they all worth? When a study fails to reproduce, it casts doubt upon the quality and conclusions of existing research. A more critical problem is that irreproducibility also seems to violate a number of related norms of science: objectivity, reliability, repeatability, robustness, generalizability, stability, universal application, and predictability. This explains why reproducibility is such an important matter in the metascientific debate.

In recent years science journals have published extensively on this issue (see for instance *Nature*'s *Specials Archive* 'Challenges in irreproducible research'). The reason for this is that a large proportion of published research, especially in the applied sciences, fails to reproduce. In biomedical research, for instance, irreproducibility has been estimated to be somewhere between 51 per cent and 90 per cent (Freedman et al. 2015, Begley and Ioannidis 2015), and in attempts to replicate fifty-three of what are considered landmark studies in cancer biology, only six of them came out positive (Begley and Ellis 2012, see also Freedman and Inglese 2014). But this problem is not restricted to lab research. In economics, a team of researchers managed to replicate the results of twenty-nine out of fifty-nine papers from top journals (Chang and Li 2015), while Camerer and colleagues (2016) successfully reproduced eleven out of eighteen studies published in *American Economic Review* and the *Quarterly Journal of Economics*. The largest reproducibility project, however, comes from psychology. In this study, ninety different teams of researchers attempted to reproduce the results from ninety-eight published papers, of which only 39 per cent gave a positive result (Open Science Collaboration 2015). These numbers are themselves debated. One could ask how representative the samples are, for instance, and whether the problem lies in the original or in the repeated study. But the general conclusion remains: something is not right.

In a survey carried out by *Nature* in 2016, 90 per cent of the scientists who responded agreed that there is a reproducibility crisis in science, with 58 per cent characterizing the crisis as major (Baker 2016). Suggestions of how to solve the problem of irreproducibility usually focus on limitations and flaws within the scientific community, such as lack of transparency, cognitive biases, statistical tools, misconduct, errors, publication pressure, and insufficient data. While these are important issues to raise, with potential for improvement, it might not be the full story. Is the scientist always to blame when results fail to reproduce? Or could we also challenge the principle of reproducibility itself? After all, the expectation that a study can be performed the same way twice and deliver exactly the same result is one that is heavily laden with philosophical assumptions, especially about the nature of causation.

In this chapter, we address the philosophical basis for the principle of reproducibility. By doing so, we hope to explain why the way we understand reproducibility must be reviewed in light of the insights from previous chapters. In its current form, this principle might be the best example of how much the orthodox understanding of causation, challenged throughout this book, has actually influenced science.

28.2 Four Assumptions about Causation

There are different ways to interpret the principle of reproducibility. In one version, reproducibility means that the study is repeatable, including the results. Often such repetitions will anyway be carried out by the researchers before publication to ensure that the result can be trusted and was not just a one-off event. This might not be what is usually meant by reproducibility, since the aim here is that the research can be repeated elsewhere, by a different researcher or research team. We might then instead speak of 'replication' or 'duplication' of a study. That a study is replicable typically means that someone else can repeat the study and obtain the same results, and that the replicated study should be as similar as possible to the original one (Casadevall and Fang 2010). Transparency is then an important criterion, since one will need to have detailed information about the original study in order to replicate it: study design, data, tools, sample, and so on. If reproducibility is understood in this way, then what we are trying to do is to recreate the situation in which the original results were produced.

When a study is repeated or replicated, then the way in which the results are obtained is as important as what exactly those results are, and perhaps more so. But reproducibility seems to also carry with it the expectation that the results are the same, or at least similar, so that the conclusions of the study confirm those of the original study. After all, would we say that reproducibility was established unless the results are in line? One could think of a case where a study has been repeated but where the results diverge. We then have two common responses: (1) that there was a causally relevant difference between the two studies, or (2) that the results of one of the studies cannot be trusted. But a third reply is possible: (3) causation does not work like that.

The principle of reproducibility, at least as stated above, is motivated by some very concrete expectations about the nature of causation. When we try to repeat as much as possible from the original study in the attempt to produce sufficiently similar results, we are assuming that *the same cause, under the same condition, should give the same effect.* But, if this is a requirement for reproducibility, then only certain types of research could possibly meet it. For instance, we would need to deal with a closed or isolated system, in which all causal inter-actions and events are determined by a finite set of initial conditions of which we have a complete overview. This might work when dealing with theoretical or abstract models but nothing in the real world works this way, even in controlled lab settings.

From the expectation that the same cause should always give the same effect, then whenever we get a different effect we can conclude that something must have been different in the cause. This is why we often speak of ideal conditions, which usually just mean the conditions under which the effect is successfully produced. A further assumption about causation is that *a cause must necessitate or guarantee its effect.* If the causal set-up of the study is right, then the results must follow, assuming that we have what Mill (1843: 217) called a *total cause.* When replicating a study it is essential that all the causally relevant factors are the same. If not, the same results cannot be expected. But this only holds if we are working within a *deterministic* and *closed system* in which the effect can be predicted with absolute certainty given a precise and complete set of initial conditions. If a chancy, chaotic, or indeterministic element is included, the result cannot be predicted or guaranteed.

To sum up, the principle of reproducibility rests on four assumptions about causation, all of which relate to an empiricist conception. These assumptions are not restricted to the norm that research should be reproducible, but underlie most of scientific methodology.

1. Same cause gives same effect.
2. Causal necessitation.
3. Total cause.
4. Deterministic and closed system.

In this book, we have offered philosophical concerns about all four assumptions. Now it seems that the reproducibility crisis in science arises at least partly because this concept of causation is a poor match for reality. Instead, science has to deal with real causation occurring in *open systems*, with many *unknown or uncertain factors*, interacting through *non-linear processes*, containing *chancy or hypersensitive elements*. In such a context, a miniscule change can significantly alter the outcome. This reality is what most scientists have to deal with and what their results and predictions are being evaluated against. Might it be that science is being held to an unrealistic ideal? Perhaps some theoretical or computer-simulated research could match the assumptions 1–4 above. But how about everything else?

Questioning the principle of reproducibility as a scientific criterion might seem like a risky business. Wouldn't that open up the door for all kinds of poor-quality and idiosyncratic research? It could, but not necessarily. There are versions of reproducibility worth having—that also preserve the aim of objectivity and generalizability of research. Before we go on to consider such alternatives, however, we will return one last time to a discussion that has been an ongoing theme of this book: the problem of causal interference.

28.3 Reproducing the Same or Similar What?

The issue of causal interference goes to the heart of the norm of reproducibility. As we have seen, reproducibility depends on the possibility of an independent researcher replicating an experiment or study, and its outcome, under the same conditions as the original study. But what exactly counts as the same conditions? Is it sufficient that the conditions are similar? And would similarity of conditions guarantee similarity of results?

In Chapter 8 we saw how causal interference and prevention are traditionally taken as problems for causation, to be explained away and eliminated. Two strategies were discussed for doing so: causal isolation and causal expansion. The first strategy is to include more in the cause, while the second is to isolate the cause from possible interferers. In the attempt to preserve reproducibility, both strategies are used and both have their problems.

If we want to make sure that the same (or similar) result is obtained whenever a study is repeated, we need to control and limit contextual factors. Isolation, idealizations, and the use of models is one way to keep interferers to a minimum. This again increases the chance of causal regularity and robustness, thereby promoting reproducibility of the result. The lab setting is designed to mimic a closed system not unlike that of the model. For those who do not work in a lab, much can be done with the study design or in data analysis to isolate the cause from other factors.

Nevertheless, even in lab settings where context is limited and controlled, a very small difference in the experimental set-up might produce a vast difference in outcome: changes in temperature, light, humidity, shaking, or stirring. The possible interferers will vary from case to case. In Chapter 26 we saw how even the medium in which a cell was cultivated made a significant difference to the outcome, something that was not previously known. What can we then expect when dealing with the complexities of ecological systems, animals, humans, or societies?

This is why it is so difficult to know which parts of a study one is supposed to include in the replication since only what is relevant for the outcome needs to be replicated. Any causally irrelevant factor need not be the same, but what counts as causally relevant is exactly that which can causally affect the result. As noted by Drummond (2009), we might say that what we aim to replicate are the results, not all the details of the study. But this leaves the question of what counts as successful replication purely a matter of creating the same result under similar enough conditions, where 'similar enough' just means that we got the same results.

For replication of the study, if we don't know which factors are causally relevant to the outcome, then everything is potentially equally important. The best way to test reproducibility would then be by perfect and complete replication. For an experiment to be reproducible, then, it has to come with a very detailed lab protocol, specifying everything that might play a causal role. This is a form of causal expansion, and has been suggested as a strategy to solve the irreproducibility crisis. An editorial in *Science* calls for more transparency to increase reproducibility, urging scientists to describe their experiments in more detail: 'An example for animal experiments is reporting the source, species, strain, sex, age, husbandry, inbred and strain characteristics, or transgenic animals, etc. For cell lines, one might report the source, authentication, and mycoplasma contamination status' (McNutt 2014). Other science journal editors have emphasized that also facts about the lab environment should be reported, since these can affect the outcome of an experiment, as argued in 'Chow down' (2016). This article refers to a study by Begenisic et al. (2015), who found that female mice who had access to wheels, stairs, and tunnels interacted differently with their offspring, which then positively affected their cognitive development. If this study were to be repeated without the same possibility for play and exercise, one should not expect the results to be the same.

Light, heat, food, company, exercise, distractions, stress—all are at the fingertips of scientists who set up mouse experiments. Subtle changes in any of these can lead to profound, and potentially useful, discoveries about how health is changed by external factors ... Given that we know that environment affects the outcome of experiments, it is surprising that we don't know more about the environmental set-up of other studies—those that test the impact of a potential medical treatment, for example.   ('Chow down' 2016)

Transparency is not only important in the lab. When using computer modelling and simulations in the analysis of data, for instance, one might have to share the full code and data sample in order to make sure that the study can be replicated perfectly. In other research, however, it might not be possible or even desired to use the same sample when replicating a study. While we agree that it is important for the sake of theory development to understand how contextual factors can causally affect the result, it would not be a good idea to try to replicate all these conditions when testing a theory for reproducibility. It could even have the opposite effect. The more detailed the protocol, the smaller the chance of repetition. After all, one cannot replicate everything in a study. Drummond (2009) makes a stronger claim. Instead of understanding reproducibility as being demonstrated through replication, he thinks perfect replication holds very little power and is a poor substitute for real reproducibility. If what we are interested in is the robustness and generalizability of a theory, it is more important to see whether the results can be reproduced in different ways than in the original study.

We have argued in this book that it is a mark of causation that it is vulnerable to interference. Scientifically, this need not be a problem but can instead be treated as an asset. In Chapter 26 we showed how the possibility of causal failure can actually help us in developing theories. This can be achieved by testing hypotheses under different conditions. A detailed protocol could be useful in this respect since it allows us to compare set-ups and study the differences, but as a way to increase the chance of getting the same results for the purpose of reproducibility, it seems counterproductive. Not only will it make repetition of a study practically impossible; it will also carry less epistemic force. Understood in this sense, reproducibility seems to work best if what we replicated were models, not real-life events.

What happens within an idealized context might not have much to do with the real-life phenomenon that it was intended to represent. When we fail to get the results predicted from the idealized model, we might conclude that if only reality were more like the idealization, the result should have followed. What, then, does this teach us about real causation?

28.4 Reproducibility Reconsidered

Is there a way to keep reproducibility as a more realistic scientific ideal? Absolutely. But then reproducibility must mean something different from the mere replication of studies and results. If a scientific result can only be obtained using a certain strain of mice from a single producer being fed and exercised in exactly the same way, then how significant are the results that we gain from the studies? And if the results of a behavioural study only applies to WEIRD samples (Henrich et al. 2010) or a business strategy only works in a single company, then we have lost the generalizability of research that reproducibility was supposed to contribute. What we gain in precision we lose in application.

Instead of making more detailed study protocols, then, in order to increase reproducibility, one way to make the results more robust is by focusing less on reproducing outcomes using the exact same models and methods. Drummond (2009) suggests that the only type of reproducibility worth having is one that requires change, unlike replication, which is an attempt to avoid change. If a result can be trusted, it should be possible to arrive at it in different ways, with different samples and perhaps even via different methods. It should be noted, however, that it cannot be the precise results that are to be reproduced, but rather the causal conclusions and insights that the results indicate.

Casadevall and Fang (2010) argue that the distinction proposed by Drummond between replicability and reproducibility is a useful one. Among other things, the distinction reveals that scientists are not generally interested in the possibility of precise replication, which might anyway only be obtainable by using the same lab and same experimenter, but rather in results that contribute to the understanding of a phenomenon. 'When findings are so dependent on precise experimental conditions that replicability is needed for reproducibility, the result may be idiosyncratic and less important than a phenomenon that can be reproduced by a variety of independent, nonidentical approaches' (Casadevall and Fang 2010: 4973). Different approaches supporting the same causal conclusion carry more epistemic weight than replication of a study. One should not need to use a particular sample or species to demonstrate evolutionary processes, or planets to demonstrate gravitational force.

28.5 Thinking outside the Idealized Box

The reproducibility crisis in science tells us something. Some say that it mainly tells us something about science, for instance that most scientific findings are false (Ioannidis 2005), that scientists are careless, biased, and fraudulent, or that some disciplines are simply studying too high a level of complexity to count as real science. But there is another possibility. Perhaps it shows that the Platonic-Galilean ideal of science is unsuited to the messy reality against which the insights of applied sciences are tested and evaluated. If so, it is time to replace the norms of ideal science with a new set of norms for doing science in the real world; for real people, real situations, real organisms, and with a realistic standard for causal prediction.

To turn this around is not an easy task and cannot be done without support from the scientific establishment, such as journal editors and funding institutions. No norm of science exists in a vacuum, but develops over time and across disciplines. For a start, we need to stop thinking inside the box of the idealized model where context, complexity, and variation are enemies of causal knowledge.

During the course of this book, we have tried to show that causation is quite different from what Hume described, yet scientific methods and norms rest to an extent upon Hume's philosophical legacy. What could differ if we changed the way in which we understand causation? Potentially, a lot. In this chapter we have argued that perfect replicability should not be an aim for science. The idea that same cause gives same effect is one of the most deep-seated ideas of causation, much like the idea of replicability of research. A dispositional theory rejects this assumption on meta-physical grounds. Instead of producing perfect regularities, causation produces tendencies. So rather than expecting that a cause invariably or necessarily produces its effect under certain but unspecified ideal conditions, we should search for something that disposes or tends toward the effect. Such tendencies will not be found in isolation but only when the cause interacts with other factors. This is because causation is essentially complex.

To study the causal powers of things by keeping them isolated from other factors, therefore, might not be the best approach. If we only look at what happens within an unnatural or ideal context, involving sterile environments and no interferers, we will miss out on valuable causal knowledge in our attempt to pin it down. Once we move from the model or the lab to a real-life setting, then, predictability diminishes considerably. Especially if the predictions are based on what happens under the condition of the idealized context.

Within the medical field, it is taken as given that lab results cannot be directly transferred beyond the lab. This is why there are three steps: pre-clinical studies with physiological and animal studies, followed by clinical studies on humans, such as randomized controlled trials, and, finally, post-market follow-up studies in persons while they are under treatment. Starting from the theoretical model with minimal complexity, each step introduces more complexity than the previous one. Could the same be done in other disciplines? And is it possible to include more variation and complexity in otherwise controlled studies? Some scientists have promoted such ideas (e.g. Boysen et al. 2011, Suryanarayanan 2013), but they are far from mainstream. One reason for this is that such proposals clash so obviously with other established norms of science.

Scientific methodology has sought, since Plato at least, to extract universal truths and regular behaviour from the messy reality. By removing ourselves from complexity and context, we have tried to find the core, essence, or real nature of things. This essence seems to exist only in its pure form within theoretical models, abstractions, isolations, perhaps even thought experiments. But this is a theoretical construct, not the reality with which we started. It also misses out on crucial elements of scientific discovery discussed in Part VIII: that science is a hands-on, dynamic activity, in which the full causal potential must be teased out by considering different contexts, different methods, and even cases of causal failure.

Conclusion: New Norms
of Science

At the beginning of the book we explained how science is a normative enterprise. We considered some of the *prima facie* norms of science; that science should be objective, for instance. Having now reached the end of our investigation into causal science—the science that underpins our discovery and understanding of causal connections—we are in a position to propose a set of norms for the correct conduct of that significant branch of science. We do not claim that this set of norms is exhaustive. It is unlikely that any such list could be made. But they are norms that we think come out of the foregoing discussions and perhaps delineate what is distinctive about our view. We explain each norm, recapitulating the evidence in its favour, and identify the most significant chapters of the book in which the arguments can be found. After that, we finish with a summary of each norm.

The norms are:

1. The metaphysics norm
2. The causal norm
3. The norm of involvement
4. The tendency norm
5. The norm of deep understanding
6. The norm of negative results
7. The symptoms norm
8. The fallible norm
9. The contextual norm.

Metaphysics (chs 1, 2, 11, 24, 28)

This is the norm we offer of the highest level of generality. Indeed, the metaphysics norm applies not just to the part of science that we are calling causal science but to all of science. None of it is metaphysics-free. In the specific case of causation, however, there is a special importance in this norm because of the way metaphysics relates to methods. The view we have established is that metaphysical commitments are reflected in the methods that we adopt for the discovery of causal connections.

One must first have a view about what a phenomenon is before one can look for it. If one thought that causation was nothing more than regularity—a pattern of constant conjunction that can be found to hold between certain types of events but not others—then it would be perfectly reasonable to adopt a method that identified those regularities. One would then have found causation. We have shown how different approaches to finding causation each seem to commit to a view of what it is. Some methods search for differences that are made, for example, which displays commitment to the view that causes are difference-makers.

The superficial reason why a method of investigation is adopted may simply be that a scientist has been trained to use the method, as in the case of epidemiology. The epidemiologist may say that he or she has no view of what causation is and is not making causal claims. However, we have shown how there is another kind of reason why a method is adopted and developed. This is because it is thought to be a good way of discovering causes, which means that it is believed to capture some aspect of the true nature of causation. It follows that the use of a method reveals one's metaphysical commitments regarding the nature of causation, be they only implicit. To quote E. J. Lowe, 'We are all metaphysicians whether we know it or not, and whether we like it or not' (Lowe 2002: 4). Methodological approaches to causation will reveal those commitments even if we do not acknowledge them explicitly.

We shouldn't ignore metaphysical commitments, therefore. Indeed, it makes sense to understand exactly what they are and, preferably, to be able to reason about them and decide the best position for the metaphysics of causation. One could then choose the methods for discovering and understanding causation that best fit the reality of the phenomena. This is not to say, however, that the metaphysics of causation cannot be informed by an empirical understanding of the details of real causal processes. As with Neurath's boat, we can work on causation in both an abstract, philosophical way and a concrete, scientific way simultaneously, ideally converging on a theory that is satisfactory on both counts.

We should also acknowledge that metaphysical matters are contested. There is no consensus on the one true theory of causation, though the same is said of almost every other domain of human enquiry. We have made a case for our preferred theory, arguing that it makes sense of some scientific practices but might require a revision to others. This is unlikely to end the debate. But the norm we advocate concerns the necessity of metaphysical awareness, even in the empirical areas of knowledge that seem non-philosophical. It is always best to know one's metaphysical commitments and then be able to defend them.

Causation (chs 2, 11, 12, 18, 21, 25)

The causal norm is the one most central to the aims of this book. The norm just tells us to accept that causal matters are indispensable to our scientific understanding of reality. Science should not avoid drawing causal conclusions. An attempt to do so would be mere pretence. If one thinks that an RCT is not about trying to establish a causal connection, for instance, then there really seems to be no motivation for running it. Science aims to be useful: to have an application. One way in which it can do this is by allowing us to change the world, or perhaps to change ourselves so that we treat the world better. It is hard to see how this will be possible, or make any sense, unless causation is accepted as real. It is quite appropriate that in science we seek causal truths, to use causal truths, and premise the notion of empirical observation upon causal truths.

Of course, one is not to assert causal claims without warrant. If the discovery of causal connections is important, it is also important to know when there are none. Much of the philosophy of causal science concerns precisely when there is such warrant for a causal claim. But a reluctance ever to acknowledge that the business of science is to discover causes and use them, or to at least acknowledge that this is a legitimate scientific goal, is mistaken, in our view, and quite possibly self-undermining. Science should not avoid drawing causal conclusions but it should avoid drawing them prematurely. The question of appropriate methods in the causal sciences concerns precisely the issue of when causal conclusions are warranted and when they are not.

Involvement (chs 2, 8, 11, 12, 24, 28)

We must accept that we are causally involved with the world and, indeed, that this is a precondition for the possibility of knowledge. Unless we are causally connected with the world, we are incapable of knowing it. We cannot even make observations of it. This allows us to distinguish the issue of objectivity versus subjectivity from the realism and anti-realism divide.

Objectivity and realism are typically seen as allied, as are subjectivity and anti-realism. Realism is thought to require no point of view on the world. The subject of experience and knowledge should be kept out of the account. However, the norm of involvement is an acknowledgement that we must be in a (causal) interaction with the world in order to know it. In that case, realism actually requires a point of view on the world. This also shows why idealization fails us: attempting a view on the world as if from nowhere, abstracting away from the messy reality. But we are a part of that messy reality, both in terms of our intentions, our interventions, our purposes, and our observations.

A simple norm of objectivity will not suffice, for it presents us with an unobtainable ideal. We choose which observations to make, which experiments to perform, which causal connections are important for us to know, and to what use we will put such knowledge. Utility of knowledge will often be a driver for its acquisition. All these points tell us that we are involved in the world we investigate. A world in which we are uninvolved, such as the supernatural realm, is for that reason unknowable.

But there is nevertheless room for a realist conception of the world and of knowledge. The interactions that we have are real enough. They are a basis for knowledge. And there are scientific methods at our disposal that allow us to build a justified theory of the world. Furthermore, the norms for the correct scientific acquisition of knowledge are open to rational consideration and challenge. Knowledge is neither easy to gain nor certain, yet there are sufficient grounds for resisting relativism.

Tendencies (chs 8, 9, 10, 17, 19, 20, 21, 28)

We come now to a norm that emerges from our own particular view of causation, whereas the preceding norms could be held by anyone who took the reality of causation seriously. However, it is not necessary to accept every detail of our account in order to accept this norm, nor to accept the entire metaphysics of causal powers or dispositions, even though we think of that as the best way to ground the norm. The norm is based simply on the empirically justified claim that repetitions reveal tendencies. The norm tells us, then, that in searching for causes, it is adequate that we look for tendencies or trends rather than perfect regularities. The latter, as we have argued, would require idealization or abstraction. Indeed, perfect regularities found in nature, if there are any, are more likely indicative of something else that is not causation, such as identity, or a characterizing feature. Mortality always accompanies humanity, for example, but not because one causes the other.

We used a notion of less than perfect regularity and allied notions such as raised incidence to characterize the data that can be gathered in evidence of causal connections. But we also gave notes of caution. Such regular behaviour need not always be available as it requires there to be multiple instances of the cause. And the frequency of occurrence of an effect following the cause, even where there are a large number of instances, might not reflect the true strength of the cause. We have to be content, then, with a degree of epistemic humility. Despite knowing all the data, we can still be wrong about some of our causal assessments.

But we think that a tendency view makes much sense of how science is conducted. It explains why we make tentative judgements, for instance, often qualified in terms of probability. Predictions will be fallible, theories will be developed over time, our confidence increasing as more evidence is acquired, but such data will never strictly entail the truth of the theory. We also saw that with such a theory we are in a position to understand a more naturalistic account of probability: one that is less mathematical, perhaps, and also applicable to degree of belief. Such an account explains how there can be more than enough reason to believe a theory while its classical chance is still less than 1.

Deep Understanding
(chs 7, 11, 13, 14, 16, 23, 24, 25, 26)

The norm of deep understanding tells us that causal science should also be about understanding causation and not just discovering it. These two intellectual endeavours are not mutually exclusive in that deep understanding is one way in which a causal connection is discovered. But we advocate the norm because one could know that A is causally connected to B without knowing how. The former knowledge can be useful, of course, but it is also incomplete, which may limit what we are then able to do with such knowledge.

Semmelweis may have discovered a causal connection between washing of hands and the lower mortality rate of his patients but he didn't understand it. He didn't know the mechanism or what differences would be made by variations on his routine or what could stop it working. There is of course more than one possible way in which a cause can get from A to B. In some cases, there is more than one actual way. Heat transference can be by convection, conduction, or radiation and there will be instances where it is important to know which of these processes is at work.

Deep causal understanding comes when we have a rich theory: one that tells us not just what causes what but also how or why. There are at least two main reasons why this is preferable. One is that a richer theory enables us to reason counterfactually about a variety of interventions and changes. Semmelweis only knew that doing what he did resulted in a lower mortality rate. What alterations could he have made to his routine, if any, without changing that result? Were some of his actions unnecessary? Of course, he could have just made a change and looked at what followed. But people might have died. If we can get into a position where we have good reason in advance to think that an intervention will be beneficial rather than harmful, then it is often preferable to trial and error. A rich theory allows you to understand the mechanisms, the context sensitivities, actual and possible failures, it facilitates non-monotonic reasoning, and so on.

A second reason why a rich causal theory is preferable is simply that it gives us increased confidence in the theory. Knowing how A causes B supports the theory that A does cause B, rather than the connection between A and B being a mere association, for example. It is preferable that a causal conclusion is linked to a well-developed theory and one that is backed by a variety of evidence. This is why we are suspicious of a strict evidence hierarchy that has randomized controlled trials, and their systematic review, outweighing any other form of evidence.

Negative Results (chs 5, 8, 12, 26)

The norm of negative results tells us that there are two sides to causal discovery. Repeatedly corroborating a theory involves a diminishing return. Beyond a point, each new corroboration adds less to our confidence in the truth of the theory.

Negative results—cases of causal failure—contribute significantly to the understanding of the theory. They provide an opportunity for the growth and development of the theory. All causal tendencies will be context-sensitive, for instance, susceptible to counteraction. Causal failure reveals how changing contexts can block a typical outcome and produce a different one.

In the spirit of Popper's falsificationism, then, the norm tells us that deep causal understanding is developed as a result not just of trying to prove a theory right, but also trying to prove it wrong; or at least of testing the theory's limits and boundaries. Such failures indicate where theories are to be nuanced and further expanded.

The norm also trades on a further obvious point. Causal science will be badly served if we concentrate only on the corroboration and ignore the counterexamples, the discrepancies. There is a clear methodological flaw if we are interested only in the cases where A is followed by B, and draw a causal conclusion on that basis, while ignoring cases of A that are not followed by B. Allied to this is the point that, for any two variables, it can be just as important to know that one doesn't cause the other as to know that it does.

Symptoms (chs 3, 4, 15, 20, 23, 27)

Success under a method should be treated as no more than a symptom of causation, according to the norm of symptoms. Furthermore, one should be open to evidence acquired through plural methods because causation has plural symptoms. This is what we call the symptomatic approach.

The norm may require a revision to some ways in which evidence is understood. According to our account, with its metaphysical primitivism about causation, success under a method cannot straightforwardly equate to the truth of a causal claim. A method does not put us in a position to pronounce that, yes, A causes B; or, no, it doesn't. Each method will itself contain many normative elements, for the correct experimental procedure to follow, and for when a positive result is to be declared. But, according to the symptomatic approach, there will also be the possibility that a method declares success when there is no causal connection involved, and vice versa. Hence, the symptomatic approach is the opposite of operationalism, which takes success under the method as definitive of the phenomenon in question. We take it that good methods can be indicative of causal connections, which is what makes them good. But, given that they all identify something that is symptomatic of causation only, rather than causation itself, then it is still possible for a good method to produce the wrong result. The more types of evidence, supporting the same causal conclusion, the better. And even though we can have more than enough reason to believe a theory, we have to admit the possibility that it could still be false. The norm also makes it clear why it is so important, of course, to identify the correct symptoms of causation, through a good metaphysical investigation into the nature of causation itself.

Fallibilism (chs 8, 17, 19, 22, 27)

Even our best theories and predictions are fallible. This is true of science generally but of causal theories in particular. However, absolute certainty is not required for rational belief in a theory. There can be good reasons to believe in a theory even if it still might be false. Methods of identifying and understanding causal connections are good when they tend to get those causal claims right. And one method is better than another if it tends more to get it right. But a theory can be based on the best evidence and still be false, which is a view we call epistemic humility. The fallible norm tells us that we ought to believe the theories for which there is good evidence despite the possibility of them being false.

This of course means that we have to live with a number of leftover philosophical problems. It is not easy to be humble. The fallible norm gives us a mix of anti-realism and realism that could annoy both sides of that debate. Fallibilism displeases the realist by allowing that full knowledge of all the data still does not guarantee truth of a theory. But it displeases the anti-realist by accepting that there are real causal facts that lie beyond our theories and which we still might be unable to know. The anti-realist could find this particularly troubling, for how do we ever know that the real facts of causation match or mismatch with our best theories? The answer, of course, is that we don't. Those metaphysical facts of causation are a theoretical postulate but, as we have argued, one that there is good reason to accept. To put the norm another way, causal knowledge does not have to be infallible, and nor do good predictions.

Context (chs 5, 6, 7, 8, 23, 28)

The context norm urges us to embrace the messiness. But what does that mean and why should we do it? As we have argued, philosophically and methodologically, it looks as though we have tried in the causal sciences to abstract away from the world's complexity. We have variously screened off interferers, modelled causes in simulations, isolated test subjects, produced identical test subjects (such as mice models), considered thought experiments, constructed laws that apply only *ceteris paribus*, and so on. Now these approaches all have their uses and are able to provide conclusions that have some value, such as discovering the natural tendencies of causes. But, as we have emphasized throughout the book, such approaches also impose a number of limitations on our science. What of external validity, for example? How can we be so sure that instances outside the laboratory will resemble those within the laboratory? And how will the methods of discovering causes allow us to subsequently put them to use? Furthermore, are we right to assume that all of simplification, abstraction, and idealization are harmless intellectual crutches? What if they were to naturally support a monocausal model, when in reality almost every cause is complex? What if we think we know the effect of each causal factor in isolation and simply treat complex cases as if they will be the mere aggregation of those effects? That would not allow for non-linear and emergent interactions. Stated so bluntly, the idea may sound obviously problematic and yet we do find in both methods and policy that we are expected to reason that way. If A, B, C, and D are all considered safe, a default assumption is that the combination and A, B, C, and D is also safe.

Ours is indeed a messy world with many different processes under way, often interfering with each other's natural passage, and with unpredictable and inde-terministic events occasionally mixed in. Composition is not always additive. We sometimes get new and interesting phenomena that emerge from the combination of more simple elements. We certainly need a philosophical theory of causation that fits such a world; and corresponding methods, able to produce useful theories that tend to be true.

Here is a final summary of our proposals:

1. Metaphysical commitments concerning the nature of causation should be acknowledged and critically examined. To adopt a method of investigation is to accept, if only tacitly, the metaphysics that justifies that method; there is thus no point in adopting methods unless they reflect the true nature of causation. (*The metaphysics norm*)
2. Causation is vital in science, in terms of its foundations and utility, and its investigation should be seen as a core activity. Science should embrace the discovery of causal connections as a worthy goal. While careful to apportion our beliefs to the evidence, the discovery and utilization of causal connections is rightly to be considered one of the central aims of science. (*The causal norm*)
3. We must accept that we are causally involved with the world and, indeed, that this is a precondition for the possibility of knowledge. Seeking scientific methods that have no role for the practitioners and users of science is not a worthy goal in itself. (*The norm of involvement*)
4. We should take tendencies as indicative of causal connections. We should expect different effects in different contexts, rather than perfect regularities or necessary connections. (*The tendency norm*)
5. Causal science should be about understanding causation and not just discover-ing it. We should aim for rich theories that tell us not just what causes what but also how and why. (*The norm of deep understanding*)
6. Our causal theories should be developed also in line with negative results—causal failure—rather than merely repeated corroborations. Discrepancies present a major opportunity for new knowledge. (*The norm of negative results*)
7. Treat success under a method as symptomatic, not definitive, of causation. Adopt plural methods, those that best reflect the plural symptoms of causation. (*The symptoms norm*)

8. Absolute certainty should not be expected: it is neither required nor possible for rational belief in a causal theory. Because of the primitive nature of causation, it is possible that all the available knowledge still leaves us short of certainty. (*The fallible norm*)

9. Embrace the messiness, the complexity: the real rather than the ideal. (*The contextual norm*)

References

'Challenges in Irreproducible Research', Specials Archive, *Nature*, <http://www.nature.com/news/reproducibility-1.17552>.

'Chow Down: Scientists Should Pay More Heed to the Varying Effects of Diet and Environment on Animal Work' (2016), Editorial, *Nature*, 530: 254.

'Let's Think about Cognitive Bias' (2015), Editorial, *Nature*, 526: 163.

Abelson, M. B. and McLaughlin, J. (2011) 'What Makes Someone a Non-Responder? A Look at Typical Non-Responders to Drug Therapy and the Factors that Have Made Them That Way', *Review of Ophthalmology*, 9 September, <https://www.reviewofophthalmology.com/article/what-makes-someone-a-non-responder>.

Ahn, W. K., Kalish, C. W., Medin, D. L., and Gelman, S. A. (1995) 'The Role of Covariation versus Mechanism Information in Causal Attribution', *Cognition*, 54: 299–352.

Aldrich, J. (1995) 'Correlations Genuine and Spurious in Pearson and Yule', *Statistical Science*, 10: 64–376.

Altman, N. and Krzywinski, M. (2016) 'Points of Significance. Analyzing Outliers: Influential or Nuisance?', *Nature Methods*, 13: 281–2.

Andersen, F. and Becker Arenhart, J. R. (2016) 'Metaphysics within Science: Against Radical Naturalism', *Metaphilosophy*, 47: 159–80.

Andersen, F., Anjum, R. L., and Mumford, S. (2018) 'Causation and Quantum Mechanics', in R. L. Anjum and S. Mumford (eds), *What Tends to Be: The Philosophy of Dispositional Modality*, London: Routledge.

Anjum, R. L. (2016) 'Evidence-Based or Person-Centered: An Ontological Debate', *European Journal for Person-Centered Healthcare*, 4: 421–9.

Anjum, R. L. and Mumford, S. (2017a) 'Mutual Manifestation and Martin's Two Triangles', in J. Jacobs (ed.), *Causal Powers*, Oxford: Oxford University Press, pp. 77–89.

Anjum, R. L. and Mumford, S. (2017b) 'Emergence and Demergence', in M. Paoletti and F. Orilia (eds), *Philosophical and Scientific Perspectives on Downward Causation*, London: Routledge.

Anjum, R. L. and Mumford, S. (2017c) 'A Philosophical Argument against Evidence-Based Policy', *Journal of Evaluation in Clinical Practice*, 23: 1045–50.

Anjum, R. L. and Mumford, S. (2018a) *What Tends to Be: The Philosophy of Dispositional Modality*, London: Routledge.

Anjum, R. L. and Mumford, S. (2018b) 'A Process Theory of Causation', in D. J. Nicholson and J. Dupré (eds), *Everything Flows: Towards a Processual Philosophy of Biology*, Oxford: Oxford University Press.

Anjum, R. L., Kerry, R., and Mumford, S. (2015) 'Evidence Based on What?', *Journal of Evaluation in Clinical Practice*, 21: E11–12.

Anjum, R. L., Mumford, S., and Myrstad, J. A. (2018) 'Conditional Probability from an Ontological Point of View', in R. L. Anjum and S. Mumford, *What Tends to Be: The Philosophy of Dispositional Modality*, London: Routledge.

Anscombe, F. J. (1968) 'Statistical Analysis: Outliers', in D. L. Sills (ed.), *International Encyclopedia of the Social Sciences*, New York: Macmillan and Free Press, pp. 178–81.

Anscombe, G. E. M. (1971) 'Causality and Determination', *Metaphysics and the Philosophy of Mind*, Oxford: Blackwell, 1981, pp. 133–47.

Antoniou, G. and Williams, M. A. (1997) *Nonmonotonic Reasoning*, Cambridge, MA: MIT Press.

Aristotle, *Physics*, trans. R. Hardie and R. Gaye, Oxford: Oxford University Press, 1930.

Aristotle, *Metaphysics*, trans. H. Lawson-Tancred, London: Penguin, 1998.

Armstrong, D. M. (1997) *A World of States of Affairs*, Cambridge: Cambridge University Press.

Armstrong, D. M. (1978) *A Theory of Universals*, Cambridge: Cambridge University Press.

Armstrong, D. M. (1983) *What Is a Law of Nature?* Cambridge: Cambridge University Press.

Ayer, A. J. (1947) 'Phenomenalism', *Proceedings of the Aristotelian Society*, 47: 163–96.

Ayer, A. J. (1963) 'What Is a Law of Nature?', in *The Concept of a Person and Other Essays*, London: Macmillan, pp. 209–34.

Bacon, F. (1620) *The New Organon*, F. H. Anderson (ed.), Indianapolis, IN: Bobbs-Merrill, 1960.

Bachke, M. E., Alfnes, F., and Wik, M. (2016) 'Information and Donations to Development Aid Projects', *Journal of Behavioral and Experimental Economics*, 66: 23–8.

Baker, M. (2016) '1500 Scientists Lift the Lid on Reproducibility', *Nature*, 533: 452–4.

Balshem, H., Helfand, M., Schünemann, H. J., Oxman, A. D., Kunz, R., Brozek, J., Vist, G. E., Falck-Ytter, Y., Meerpohl, J., Norris S., and Guyatt, G. H. (2011) 'GRADE Guidelines: 3. Rating the Quality of Evidence', *Journal of Clinical Epidemiology*, 64: 401–6.

Bayes, T. (1763) 'An Essay Towards Solving a Problem in the Doctrine of Chances', in R. Price (ed.), *Philosophical Transactions of the Royal Society of London*, 53: 370–418. Reprinted in R. Swinburne (ed.), *Bayes's Theorem*, Oxford: Oxford University Press, 2002, pp. 122–49.

Bechtel, W. (2011) 'Mechanism and Biological Explanation', *Philosophy of Science*, 78: 533–57.

Bechtel, W. and Abrahamsen, A. (2005) 'Explanation: A Mechanist Alternative', *Studies in History and Philosophy of Science Part C: Studies in History and Philosophy of Biological and Biomedical Sciences*, 36: 421–41.

Beebee, H. (2000) 'The Non-Governing Conception of Laws of Nature', *Philosophy and Phenomenological Research*, 61: 571–94.

Beebee, H. (2007) 'Hume on Causation: the Projectivist Interpretation', in H. Price and R. Corry (eds), *Causation, Physics, and the Constitution of Reality: Russell's Republic Revisited*, Oxford: Oxford University Press, pp. 224–49.

Begenisic, T., Sansevero, G., Baroncelli, L., Cioni, G., and Sale, A. (2015) 'Early Environmental Therapy Rescues Brain Development in a Mouse Model of Down Syndrome', *Neurobiology of Disease*, 82: 409–19.

Begley, C. G. and Ellis, L. M. (2012) 'Drug Development: Raise Standards for Preclinical Cancer Research', *Nature*, 483: 531–3.

Begley, C. G. and Ioannidis, J. P. (2015) 'Reproducibility in Science: Improving the Standard for Basic and Preclinical Research', *Circulation Research*, 116: 116–26.

Bellis, M. (2010) 'Governments Confront Drunken Violence', *Bull World Health Organ*, 88: 644–5.

Berkeley, G. (1710) *A Treatise Concerning the Principles of Human Knowledge*, Dublin: Pepyat.

Bhaskar, R. (1975) *A Realist Theory of Science*, 2nd edn, London: Verso, 2008.

Bird, A. (2007) *Nature's Metaphysics*, Oxford: Oxford University Press.

Bohm, D. (1957) *Causality and Chance in Modern Physics*, London: Routledge.

Bohr, N. (1937) 'Causality and Complementarity', in J. Faye and J. Folse (eds), *The Philosophical Writings of Niels Bohr*, vol. 4, Woodbridge: Ox Bow Press, 1998, pp. 83–92.

Bohr, N. (1938) 'The Causality Problem in Atomic Physics', in J. Faye and J. Folse (eds), *The Philosophical Writings of Niels Bohr*, vol. 4, Woodbridge: Ox Bow Press, 1998, pp. 94–121.

Bohr, N. (1948) 'On the Notions of Causality and Complementarity', in J. Faye and J. Folse (eds), *The Philosophical Writings of Niels Bohr*, vol. 4, Woodbridge: Ox Bow Press, 1998, pp. 141–9.

Bolarinwa, I. F., Orfila, C., and Morgan, M. R. (2015) 'Determination of Amygdalin in Apple Seeds, Fresh Apples and Processed Apple Juices', *Food Chemistry*, 170: 437–42.

Boyle, R. (1674) 'About the Excellency and Grounds of the Mechanical Hypothesis', in M. A. Stewart (ed.), *Selected Philosophical Papers of Robert Boyle*, Manchester: Manchester University Press, 1979, pp. 138–54.

Boysen, P., Eide, D. M., and Storset, A. K. (2011) 'Natural Killer Cells in Free-Living *Mus Musculus* Have a Primed Phenotype', *Molecular Ecology*, 20: 5103–10.

Bradford Hill, A. (1965) 'The Environment and Disease: Association or Causation?', *Proceedings of the Royal Society of Medicine*, 58: 295–300.

Bridgman, P. W. (1927) *The Logic of Modern Physics*, New York: Macmillan.

Bridgman, P. W. (1959) *The Way Things Are*, Cambridge, MA: Harvard University Press.

Broadbent, A. (2013) *Philosophy of Epidemiology*, London: Palgrave Macmillan.

Buetow, S. (2015) 'Report Clinical Sources of Heterogeneity in Meta-Analysis: A Commentary on Steurer et al. (2014). "Are Clinically Heterogeneous Studies Synthesized in Systematic Reviews?"', *Journal of Evaluation in Clinical Practice*, 3: 131–3.

Bunge, M. (1966) 'Technology as Applied Science', *Technology and Culture*, 7: 329–47.

Cairney, P. (2016) *The Politics of Evidence-Based Policy Making*, Stirling: Springer.

Camerer, C. F., Dreber, A., Forsell, E., Ho, T. H., Huber, J., Johannesson, M., Kirchler, M., Almenberg, J., Altmejd, A., Chan, T., Heikensten, E., Holzmeister, F., Imai, T., Isaksson, S., Nave, G., Pfeiffer, T., Razen, M., and Wu, H. (2016) 'Evaluating Replicability of Laboratory Experiments in Economics', *Science*, 351: 1433–6.

Cartwright, N. (1979) 'Causal Laws and Effective Strategies', *Nous*, 13: 419–37.

Cartwright, N. (1983) *How the Laws of Physics Lie*, Oxford: Clarendon Press.

Cartwright, N. (1989) *Nature's Capacities and Their Measurement*, Oxford: Oxford University Press.

Cartwright, N. (1999) *The Dappled World: A Study of the Boundaries of Science*, Cambridge: Cambridge University Press.

Cartwright, N. (2004) 'Causation: One Word, Many Things', *Philosophy of Science*, 71: 805–20.

Cartwright, N. (2007a) *Hunting Causes and Using Them: Approaches in Philosophy and Economics*, Cambridge: Cambridge University Press.

Cartwright, N. (2007b) 'Are RCTs the Gold Standard?', *BioSocieties*, 2: 11–20.

Cartwright, N. (2010) 'What are Randomised Controlled Trials Good for?', *Philosophical Studies*, 147: 59–70.

Cartwright, N. (2011) 'A Philosopher's View of the Long Road from RCTs to Effectiveness', *Lancet*, 377: 1400–1.

Cartwright, N. (2012) 'Presidential Address: Will This Policy Work for You? Predicting Effectiveness Better: How Philosophy Helps', *Philosophy of Science*, 79: 973–89.

Cartwright, N. and Hardie, J. (2012) *Evidence-Based Policy: A Practical Guide to Doing It Better*, Oxford: Oxford University Press.

Casadevall, A. and Fang, F. C. (2010) 'Reproducible Science', *Infection and Immunity*, 78: 4972–5.

Chalmers, D. (2006) 'Strong and Weak Emergence', in P. Clayton and P. Davies (eds), *The Re-Emergence of Emergence*, Oxford: Oxford University Press, pp. 244–54.

Chang, A. C. and Li, P. (2015) 'Is Economics Research Replicable? Sixty Published Papers from Thirteen Journals Say "Usually Not"', *Finance and Economics Discussion Series 2015-083*, Washington: Board of Governors of the Federal Reserve System, doi:10.17016/FEDS.2015.083.

Chang, H. (2009) 'Operationalism', *Stanford Encyclopedia of Philosophy* (Fall), Edward N. Zalta (ed.), <http://plato.stanford.edu/archives/fall2009/entries/operationalism/>.

Chisholm, R. M. (1946) 'The Contrary-to-Fact Conditional', *Mind*, 55: 289–307.

Close, F. (2009) *Nothing: A Very Short Introduction*, Oxford: Oxford University Press.

Collingwood, R. (1940) *An Essay on Metaphysics*, Oxford: Clarendon Press.

Collins, J. D., Hall, E. J., and Paul, L. A. (eds) (2004) *Causation and Counterfactuals*, Cambridge, MA: MIT Press.

Copeland, S. (2017) 'On Serendipity in Science: Discovery at the Intersection of Chance and Wisdom', *Synthese*, doi:10.1007/s11229-017-1544-3.

Cousineau, D. and Chartier, S. (2015) 'Outliers Detection and Treatment: A Review', *International Journal of Psychological Research*, 3: 58–67.

Crabbe, J. C., Wahlsten, D., and Dudek, B. C. (1999) 'Genetics of Mouse Behavior: Interactions with Laboratory Environment', *Science*, 284: 1670–2.

Craver, C. and Darden, L. (2013) *In Search of Mechanisms*, Chicago, IL: University of Chicago Press.

Cross, C. (1991) 'Explanation and the Theory of Questions', *Erkenntnis*, 34: 237–60.

da Costa, L. P. and Dias, J. G. (2015) 'What Do Europeans Believe to Be the Causes of Poverty? A Multilevel Analysis of Heterogeneity within and between Countries', *Social Indicators Research*, 122: 1–20.

Daston, L. (1979) 'D'Alembert's Critique of Probability Theory', *Historia Mathematica*, 6: 259–79.

Dawkins, R. (1976) *The Selfish Gene*, Oxford: Oxford University Press.

de Finetti, B. (1974) *Theory of Probability*, vol. 1, Chichester: Wiley, 1990.

Deaton, A. (2010) 'Instruments, Randomization, and Learning about Development', *Journal of Economic Literature*, 48: 424–55.

Deaton, A. and Cartwright, N. (2018) 'Understanding and Misunderstanding Randomized Controlled Trials', *Social Science and Medicine*, doi: 10.1016/j.socscimed.2017.12.005.

Descartes, R. (1641) *Meditations on First Philosophy*, in J. Cottingham, R. Stoothoff, and D. Murdoch (eds), *The Philosophical Writings of Descartes*, vol. 2, Cambridge: Cambridge University Press, 1985, pp. 1–62.

Dowe, P. (2000) *Physical Causation*, New York: Cambridge University Press.

Dowe, P. (2001) 'A Counterfactual Theory of Prevention and "Causation" by Omission', *Australasian Journal of Philosophy*, 79: 216–26.

Dretske, F. (1977) 'Laws of Nature', *Philosophy of Science*, 44: 248–68.

Drewery, A. (2000) 'Laws, Regularities and Exceptions', *Ratio*, 13: 1–12.

Drummond, C. (2009) 'Replicability Is Not Reproducibility: Nor Is It Good Science', *Proceedings of the Evaluation Methods for Machine Learning Workshop 26th ICML*, Montreal.

Duhem, P. (1954) *The Aim and Structure of Physical Theory*, Princeton, NJ: Princeton University Press.

Dunbar, K. (2000) 'How Scientists Think in the Real World: Implications for Science Education', *Journal of Applied Developmental Psychology*, 21: 49–58.

Dunbar, K. and Fugelsang, J. (2005) 'Scientific Thinking and Reasoning', in K. J. Holyoak and R. G. Morrison (eds), *The Cambridge Handbook of Thinking and Reasoning*, Cambridge: Cambridge University Press, pp. 705–25.

Dupré, J. (1993) *The Disorder of Things: Metaphysical Foundations of the Disunity of Science*, Cambridge, MA: Harvard University Press.

Dupré, J. (2001) *Human Nature and the Limits of Science*, Oxford: Oxford University Press.

Dye, C. (2008) 'Health and Urban Living', *Science*, 319: 766–9.

Dyke, H. (2008) *Metaphysics and the Representational Fallacy*, New York: Routledge.

Earman, J. (1992) *Bayes or Bust?* Cambridge, MA: MIT Press.

Ellis, B. (2001) *Scientific Essentialism*, Cambridge: Cambridge University Press.

Erlewyn-Lajeunesse, M., Bonhoeffer, J., Ruggeberg, J. U., and Heath, P. T. (2007) 'Anaphylaxis as an Adverse Event Following Immunisation', *Journal of Clinical Pathology*, 60: 737–9.

Esfeld, M. (2010) 'Humean Metaphysics versus a Metaphysics of Powers', in G. Ernst and A. Huttemann (eds), *Time, Chance and Reduction: Philosophical Aspects of Statistical Mechanics*, Cambridge: Cambridge University Press, pp. 119–35.

Everitt, N. (1991) 'Strawson on Laws and Regularities', *Analysis*, 50: 206–8.

Eysenck, H. J. (1994) 'Meta-Analysis and Its Problems', *British Medical Journal*, 309: 789–92.

Eysenck, H. J. (1995) 'Meta-Analysis or Best-Evidence Synthesis?', *Journal of Evaluation in Clinical Practice*, 1: 29–36.

Fair, D. (1979) 'Causation and the Flow of Energy', *Erkenntnis*, 14: 219–50.

Feagin, J. (1972) 'Poverty: We Still Believe that God Helps Those Who Help Themselves', *Psychology Today*, 6: 101–10, 129.

Feser, E. (2015) *Scholastic Metaphysics*, Heusenstamm: Editiones Scholasticae.

Feyerabend, P. (1975) *Against Method*, London: New Left Books.

Feynman, R. (1967) *The Character of Physical Law*, Cambridge, MA: MIT Press.

Foulkes, M. A., Grady, C., Spong, C. Y., Bates, A., and Clayton, J. A. (2011) 'Clinical Research Enrolling Pregnant Women: A Workshop Summary', *Journal of Women's Health*, 20: 1429–32.

Freedman, L. P. and Inglese, J. (2014) 'The Increasing Urgency for Standards in Basic Biologic Research', *Cancer Research*, 74: 4024–9.

Freedman, L. P., Cockburn, I. M., and Simcoe, T. S. (2015) 'The Economics of Reproducibility in Preclinical Research', *PLoS Biology*, 13: e1002165.

Frege, G. (1879) *Begriffsschrift*, T. W. Bynam (trans.), *Conceptual Notation*, Oxford: Clarendon Press, 1972.

French, S. (2014) *The Structure of the World: Metaphysics and Representation*, Oxford: Oxford University Press.

Freud, S. (1899) *The Interpretation of Dreams*, 3rd edn, trans. A. A. Brill, New York: Macmillan Company, 1913, Bartleby.com, 2010.

Frisch, M. (2007) 'Causation, Counterfactuals, and Entropy', in H. Price and R. Corry (eds), *Causation, Physics, and the Constitution of Reality: Russell's Republic Revisited*, Oxford: Oxford University Press.

Fuller, J. (2013) 'Rationality and the Generalization of Randomized Controlled Trial Evidence', *Journal of Evaluation in Clinical Practice*, 19: 644–7.

Fuller, J. and Flores, L. J. (2015) 'The Risk GP Model: The Standard Model of Prediction in Medicine', *Studies in History and Philosophy of Science*, 54: 49–61.

Gadamer, H. G. (1975) *Truth and Method*, trans. W. Glen-Dopel, London: Sheed and Ward.

Galileo, G. (1623) *Il Saggiatore, The Assayer*, trans. S. Drake, *The Controversy of the Comets of 1618*, Philadelphia, PA: University of Pennsylvania Press, 1960.

Galileo, G. (1632) *Dialogo Dei Massimi Sistemi*, in S. Drake (ed. and trans.), *Dialogue Concerning the Two Chief World Systems, Ptolemaic and Copernican*, New York: Modern Library, 2001.

Gasking, D. (1955) 'Causation and Recipes', *Mind*, 64: 479–87.

Geach, P. T. (1961) 'Aquinas', in G. E. M. Anscombe and P. T. Geach (eds), *Three Philosophers*, Oxford: Blackwell: 65–125.

Giere, R. N. (1973) 'Objective Single-Case Probabilities and the Foundations of Statistics', in E. Nagel, P. Suppes, and A. Tarski (eds), *Logic, Methodology and Philosophy of Science*, vol. 4, New York: North-Holland, pp. 467–83.

Gillett, C. (2016) *Reduction and Emergence in Science and Philosophy*, Cambridge: Cambridge University Press.

Gillies, D. (1973) *An Objective Theory of Probability*, London: Methuen.

Gillies, D. (2000a) *Philosophical Theories of Probability*, London: Routledge.

Gillies, D. (2000b) 'Varieties of Propensity', *British Journal for Philosophy of Science*, 51: 807–35.

Gillies, D. (2011) 'The Russo-Williamson Thesis and the Question of Whether Smoking Causes Heart Disease', in P. M. Illari, F. Russo, and J. Williamson (eds), *Causality in the Sciences*, Oxford: Oxford University Press, pp. 110–25.

Glennan, S. (1996) 'Mechanisms and the Nature of Causation', *Erkenntnis*, 44: 49–71.

Glennan, S. (2002) 'Rethinking Mechanistic Explanation', *Philosophy of Science*, 69: S342–53.

Glennan, S. (2009) 'Mechanisms' in H. Beebee, C. Hitchcock, and P. Menzies (eds), *Oxford Handbook of Causation*, Oxford: Oxford University Press, pp. 315–25.

Glimcher, P. W. (2004) *Decisions, Uncertainty, and the Brain: The Science of Neuroeconomics*, Cambridge, MA: MIT Press.

Gómez-Marín, O., Prineas, R. J., and Sinaiko, A. R. (1991) 'The Sodium-Potassium Blood Pressure Trial in Children: Design, Recruitment, and Randomization: The Children and Adolescent Blood Pressure Program', *Controlled Clinical Trials*, 12: 408–23.

Goodman, N. (1947) 'The Problem of Counterfactual Conditionals', *Journal of Philosophy*, 44: 113–28.

Gower, B. (1997) *Scientific Method: An Historical and Philosophical Introduction*, London: Routledge.

Greenhalgh, T., Kostopoulou, O., and Harries, C. (2004) 'Making Decisions about Benefits and Harms of Medicines', *British Medical Journal*, 329: 47–50.

Groff, R. (2013) *Ontology Revisited: Metaphysics in Social and Political Philosophy*, London: Routledge.

Groff, R. and Greco, J. (2013) *Powers and Capacities in Philosophy: The New Aristotelianism*, New York: Routledge.

Guyatt, G. et al. (1992) 'Evidence-Based Medicine: A New Approach to Teaching the Practice of Medicine', *JAMA—Journal of American Medical Association*, 268: 2420–5.

Haidt, J. (2001) 'The Emotional Dog and Its Rational Tail: A Social Intuitionist Approach to Moral Judgment', *Psychological Review*, 108: 814–34.

Haidt, J. (2007) 'The New Synthesis in Moral Psychology', *Science*, 316: 998–1002.

Hájek, A. (2012) 'Interpretations of Probability', *Stanford Encyclopedia of Philosophy* (Winter), E. N. Zalta (ed.), <http://plato.stanford.edu/archives/win2012/entries/probability-interpret/>.

Hall, N. (1994) 'Two Concepts of Causation', in J. Collins, N. Hall, and L. Paul (eds), *Causation and Counterfactuals*, Cambridge, MA: MIT Press, 2004, pp. 225–76.

Hanfling, O. (ed.) (1981) *Essential Readings in Logical Positivism*, Oxford, Blackwell.

Hanson, N. R. (1958) *Patterns of Discovery*, London: Cambridge University Press.

Harré, R. (1972) *The Philosophies of Science*, Oxford: Oxford University Press.

Harré, R. and Madden, E. H. (1975) *Causal Powers: A Theory of Natural Necessity*, Oxford: Blackwell.

Harvard Men's Health Watch (2010) 'Marriage and Men's Health', *Harvard Health Publications* (July), Harvard Medical School, <http://www.health.harvard.edu/newsletter_article/marriage-and-mens-health>.

Hauser, M. (2006) *Moral Minds: How Nature Designed Our Universal Sense of Right and Wrong*, New York: Ecco/Harper Collins.

Hawking, S. (2011) *The Grand Design*, Random House Digital.

Haynes, L., Goldacre, B., and Torgerson, D. (2012) 'Test, Learn, Adapt: Developing Public Policy with Randomised Controlled Trials, *Cabinet Office-Behavioural Insights Team*, <https://www.gov.uk/government/publications/test-learn-adapt-developing-public-policy-with-randomised-controlled-trials>.

Hazeldine, S. (2013) *Neuro-Sell: How Neuroscience Can Power Your Sales Success*, London: Kogan Page Publishers.

Healey, R. (1992) 'Chasing Quantum Causes: How Wild Is the Goose?', *Philosophical Topics*, 20: 181–204.

Heil, J. (2004) 'Properties and Powers', in D. Zimmerman (ed.), *Oxford Studies in Metaphysics*, vol. 1, New York: Oxford University Press, pp. 223–54.

Heisenberg, W. (1959) *Physics and Philosophy: The Revolution in Modern Science*, 3rd edn, ed. R. N. Anshen, London: Ruskin House, 1971.

Hempel, C. G. (1950) 'Problems and Changes in the Empiricist Criterion of Meaning', *Review Internationale de Philosophie*, 41: 41–65.

Hempel, C. G. (1965) *Aspects of Scientific Explanation; and Other Essays in the Philosophy of Science*, New York: Free Press.

Henrich, J., Heine, S. J., and Norenzayan, A. (2010) 'The Weirdest People in the World?', *Behavioral and Brain Sciences*, 33: 61–83.

Henrique, D. S., Queiroz, C. D., Vieira, R. A. M., and Botelho, M. F. (2009) 'Energetic Efficiency of Protein and Body Fat Retention in Crossbred Bos Indicus and Bos Taurus × Bos Indicus Raised under Tropical Conditions', *Revista Brasileira de Zootecnia*, 38: 1581–6.

Hershner, S. D. and Chervin, R. D. (2014) 'Causes and Consequences of Sleepiness among College Students', *Nature and Science of Sleep*, 6: 73–84.

Hjelmesaeth, J. (2014) 'Randomised Studies: Useful for Whom?', *Tidsskrift for Den Norske Legeforening*, 19: 134.

Howick, J. (2011) *The Philosophy of Evidence-Based Medicine*, Chichester: Wiley-Blackwell.

Hume, D. (1739) *A Treatise of Human Nature*, ed. L. A. Selby-Bigge, Oxford: Clarendon Press, 1888.

Hume, D. (1740) 'Abstract of a Treatise of Human Nature', in P. Millican (ed.), *An Enquiry Concerning Human Understanding*, Oxford: Oxford University Press, 2007, pp. 133–45.

Hume, D. (1748) *An Enquiry Concerning Human Understanding*, ed. L. A. Selby-Bigge, Oxford: Clarendon Press, 1902.

Humphreys, P. (1985) 'Why Propensities Cannot Be Probabilities', *Philosophical Review*, 94: 557–70.

Illari, P. and Russo, F. (2014) *Causality: Philosophical Theory Meets Scientific Practice*, Oxford: Oxford University Press.

Illari, P. M., Russo, F., and Williamson, J. (eds) (2011) *Causality in the Sciences*, Oxford: Oxford University Press.

Illari, P. and Williamson, J. (2012) 'What Is a Mechanism? Thinking about Mechanisms across the Sciences', *European Journal of the Philosophy of Science*, 2: 119–35.

Ioannidis, J. P. A. (2005) 'Why Most Published Research Findings Are False', *PLoS Medicine*, 2: e124.

Jacobs, J. (ed.) (2017) *Causal Powers*, Oxford: Oxford University Press.

Janssen, M. and Zander, K. (2014) 'Do You Like Organic Wine? Preferences of Organic Consumers', in G. Rahmann and U. Aksoy (eds), *Proceedings of the 4th ISOFAR Scientific Conference at the Organic World Congress*, v. 2, *Building Organic Bridges*, Braunschweig: Thünen, pp. 355–8.

Kaloper, N. and Padilla, A. (2015) 'Sequestration of Vacuum Energy and the End of the Universe', *Physical Review of Letters*, 114: 101302.

Kant, I. (1781) *Critique of Pure Reason*, trans. N. Kemp Smith, London: Macmillan, 1929.

Kardan, O., Gozdyra, P., Misic, B., Moola, F., Palmer, L. J., Paus, T., and Berman, M. G. (2015) 'Neighborhood Greenspace and Health in a Large Urban Center', *Scientific Reports*, 5: 11610.

Kenny, D. A. (1979) *Correlation and Causality*, New York: Wiley.

Kerry, R., Eriksen, T. E., Lie, S. A. N., Mumford, S. D., and Anjum, R. L. (2012) 'Causation and Evidence-Based Practice: An Ontological Review', *Journal of Evaluation in Clinical Practice*, 18: 1006–12.

Kim, J. (1976) 'Events as Property Exemplifications', in *Action Theory*, Dordrecht: Springer, pp. 159–77.

Kim, J. (2006) 'Being Realistic about Emergence', in P. Clayton and P. Davies (eds), *The Re-Emergence of Emergence*, Oxford: Oxford University Press, pp. 189–202.

Kim, J. and Scialli, A. R. (2011) 'Thalidomide: The Tragedy of Birth Defects and the Effective Treatment of Disease', *Toxicological Sciences*, 122: 1–6.

Kistler, M. (2006) *Causation and Laws of Nature*, New York: Routledge.

Kolmogorov, A. N. (1933) *Grundbegriffe der Wahrscheinlichkeitrechnung, Ergebnisse Der Mathematik*, in N. Morrison (ed. and trans.), *Foundations of Probability*, New York: Chelsea Publishing Company, 1950.

Kripke, S. (1980) *Naming and Necessity*, Oxford: Blackwell.

Kroes, P. and A. Meijers (eds) (2006) 'The Dual Nature of Technical Artifacts', *Studies in History and Philosophy of Science*, 37: 1–158.

Kuhn, T. (1962) *The Structure of Scientific Revolutions*, 2nd edn, Chicago, IL: University of Chicago Press, 1970.

Kutach, D. (2002) 'The Entropy Theory of Counterfactuals', *Philosophy of Science*, 69: 82–104.

Kutach, D. (2007) 'The Physical Foundations of Causation', in H. Price and R. Corry (eds), *Causation, Physics, and the Constitution of Reality: Russell's Republic Revisited*, Oxford: Oxford University Press.

Lacey, H. (1999) *Is Science Value Free?*, London: Routledge.

Ladyman, J. and Ross, D., with Spurrett, D. and Collier, J. (2007) *Every Thing Must Go. Metaphysics Naturalized*, Oxford: Oxford University Press.

Lakatos, I. (1970) 'Falsification and the Methodology of Scientific Research Programmes', in I. Lakatos and A. Musgrave (eds), *Criticism and the Growth of Knowledge*, Cambridge: Cambridge University Press, pp. 91–196.

Landes, J., Osimani, B., and Poellinger, R. (2018) 'Epistemology of Causal Inference in Pharmacology: Towards a Framework for the Assessment of Harms', *European Journal for Philosophy of Science*, 8: 3–49.

Laplace, P. S. (1814) *A Philosophical Essay on Probabilities*, in F. W. Truscott and F. L. Emory (eds and trans.), *A Philosophical Essay on Probabilities by Pierre Simon, Marquis de Laplace*, London: Chapman and Hall, 1902.

Lewis, D. (1973a) *Counterfactuals*, Oxford: Blackwell.

Lewis, D. (1973b) 'Causation', in *Philosophical Papers II*, Oxford: Oxford University Press, 1986: 159–213.

Lewis, D. (1980) 'A Subjectivist's Guide to Objective Chance', in *Philosophical Papers II*, Oxford: Oxford University Press, pp. 83–113.

Lewis, D. (1983) 'New Work for a Theory of Universals', in *Papers in Metaphysics and Epistemology*, Cambridge: Cambridge University Press, 1999, pp. 8–55.

Lewis, D. (1986a) *On the Plurality of Worlds*, Oxford: Blackwell.

Lewis, D. (1986b) *Philosophical Papers II*, Oxford: Oxford University Press.

Lewis, D. (1986c) 'Events', in *Philosophical Papers II*, Oxford: Oxford University Press, pp. 241–69.

Lewis, D. (1994) 'Humean Supervenience Debugged', in *Papers in Metaphysics and Epistemology*, Cambridge: Cambridge University Press, 1999, pp. 224–47.

Libet, B. (1999) 'Do We Have Free Will?', *Journal of Consciousness Studies*, 6: 47–57.

Libet, B., Gleason, C. A., Wright, E. W., and Pearl, D. K. (1983) 'Time of Conscious Intention to Act in Relation to Onset of Cerebral Activity (Readiness-Potential)', *Brain*, 106: 623–42.

Libet, B., Freeman, A., and Sutherland, K. (1999) *The Volitional Brain: Towards a Neuroscience of Free Will*, Exeter: Imprint Academic.

Lie, S. A. N. (2016) *Philosophy of Nature: Rethinking Naturalness*, New York: Routledge.

Lindseth, A. (2012) 'Praktisk kunnskap', in K. J. Ims and Ø. Nystad (eds), *På Tvers: Praksiser og Teorier om Økonomi, Kultur og Natur for det Nye Årtusen*, Bodø: Universitetet i Nordland.

Lipton, R. B., Pavlovic, J. M., Haut, S. R., Grosberg, B. M., and Buse, D. C. (2014) 'Methodological Issues in Studying Trigger Factors and Premonitory Features of Migraine', *Headache: Journal of Head and Face Pain*, 54: 1661–9.

Locke, J. (1690) *An Essay Concerning Human Understanding*, P. H. Nidditch (ed.), Oxford: Oxford University Press, 1975.

Loewer, B. (2007) 'Counterfactuals and the Second Law', in H. Price and R. Corry (eds), *Causation, Physics, and the Constitution of Reality: Russell's Republic Revisited*, Oxford: Oxford University Press.

Lombard, L. (1986) *Events*, London: Routledge.

Longino, H. (2002) *The Fate of Knowledge*, Princeton, NJ: Princeton University Press.

Lorenz, E. N. (1963) 'Deterministic Nonperiodic Flow', *Journal of Atmospheric Sciences*, 20: 130–41.

Lowe, E. J. (1982) 'Laws, Dispositions and Sortal Logic', *American Philosophical Quarterly*, 19: 41–50.

Lowe, E. J. (2002) *A Survey of Metaphysics*, Oxford: Oxford University Press.

McCall, S. (1994) *A Model of the Universe*, Oxford: Oxford University Press.

MacCoun, R. and Perlmutter, S. (2015) 'Blind Analysis: Hide Results to Seek the Truth', *Nature*, 526: 187–9.

Mach, E. (1960) *The Science of Mechanics: A Critical and Historical Account of Its Development*, 6th edn, trans. T. J. McCormack, La Salle, IL: Open Court.

Machamer, P. (2004) 'Activities and Causation: The Metaphysics and Epistemology of Mechanisms', *International Studies in the Philosophy of Science*, 18: 27–39.

Machamer, P., Darden, L., and Craver, C. F. (2000) 'Thinking about Mechanisms', *Philosophy of Science*, 67: 1–25.

Mackie, J. L. (1974) *The Cement of the Universe*, Oxford: Oxford University Press.

MacNell, L., Driscoll, A., and Hunt, A. N. (2014) 'What's in a Name: Exposing Gender Bias in Student Ratings of Teaching', *Innovative Higher Education*, 40: 291–303.

Marchal, B., Westhorp, G., Wong, G., Van Belle, S., Greenhalgh, T., Kegels, G., and Pawson, R. (2013) 'Realist RCTs of Complex Interventions – an Oxymoron', *Social Science and Medicine*, 94: 124–8.

Marmodoro, A. (ed.) (2010) *The Metaphysics of Powers: Their Grounding and Their Manifestations*, London: Routledge.

Marmodoro, A. (2016) 'Dispositional Modality vis-à-vis Conditional Necessity', *Philosophical Investigations*, 39: 205–14.

Martin, C. B. (1994) 'Dispositions and Conditionals', *Philosophical Quarterly*, 44: 1–8.

Martin, C. B. (2008) *The Mind in Nature*, Oxford: Oxford University Press.

McNutt, M. (2014) 'Journals Unite for Reproducibility', *Science*, 346: 679.

Melia, J. (2003) *Modality*, Chesham: Acumen.

Mellor, D. H. (1971) *The Matter of Chance*, Cambridge: Cambridge University Press.

Mellor, D. H. (1995) *The Facts of Causation*, London: Routledge.

Mellor, D. H. (2000) 'The Semantics and Ontology of Dispositions', *Mind*, 109: 757–80.

Mellor, D. H. (2005) *Probability: A Philosophical Introduction*, London: Routledge.

Menzies, P. (2007) 'Causation in Context', in H. Price and R. Corry (eds), *Causation, Physics, and the Constitution of Reality: Russell's Republic Revisited*, Oxford: Oxford University Press, pp. 191–223.

Miles, A. (2015) 'Ongoing Problems with the E of EBM and on the Need for Person-Centered Healthcare. A Commentary on Steurer et al. (2014), "Are Clinically Heterogenous Studies Synthesized in Systematic Reviews?"', *Journal of Evaluation in Clinical Practice*, 3: 134–7.

Mill, J. S. (1843) *A System of Logic, Collected Works of John Stuart Mill*, vol. 7, Toronto: University of Toronto Press, 1973.

Mill, J. S. (1859) *On Liberty*, London: Parker.

Molnar, G. (2003) *Powers: A Study in Metaphysics*, ed. S. Mumford, Oxford: Oxford University Press.

Monton, B. and Mohler, C. (2014) 'Constructive Empiricism', *Stanford Encyclopedia of Philosophy* (Spring), Edward N. Zalta (ed.), <http://plato.stanford.edu/archives/spr2014/entries/constructive-empiricism/>.

Moore, M. S. (2009) *Causation and Responsibility*, Oxford: Oxford University Press.

Mumford, S. (1998) *Dispositions*, Oxford: Oxford University Press.

Mumford, S. (2004) *Laws in Nature*, Abingdon: Routledge.

Mumford, S. (2012) *Metaphysics: A Very Short Introduction*, Oxford: Oxford University Press.

Mumford, S. and Anjum, R. L. (2011) *Getting Causes from Powers*, Oxford: Oxford University Press.

Mumford, S. and Anjum, R. L. (2013) *Causation: A Very Short Introduction*, Oxford: Oxford University Press.

Mumford, S. and Anjum, R. L. (2015) 'Freedom and Control: On the Modality of Free Will', *American Philosophical Journal*, 52: 1–11.

Nagel, T. (1986) *The View from Nowhere*, Oxford: Oxford University Press.

Nesse, R. M. and Williams, G. C. (1996) *Why We Get Sick: The New Science of Darwinian Medicine*, San Francisco, CA: Vintage Books.

Newton, I. (1687) *Philosophiae Naturalis Principia Mathematica*, in A. Motte (ed. and trans.), *Mathematical Principles of Natural Philosophy*, 1729, revised by F. Cajori, Cambridge: Cambridge University Press, 1934.

Nicholson, D. J. and Dupré, J. (eds) (2018) *Everything Flows: Towards a Processual Philosophy of Biology*, Oxford: Oxford University Press.

Norton, J. D. (2007) 'Causation as Folk Science', in H. Price and R. Corry (eds), *Causation, Physics, and the Constitution of Reality: Russell's Republic Revisited*, Oxford: Oxford University Press.

Nuzzo, R. (2015) 'How Scientists Fool Themselves—and How They Can Stop', *Nature*, 526: 182–5.

Open Science Collaboration (2015) 'Estimating the Reproducibility of Psychological Science', *Science*, 349: aac4716.

Oppenheim, P. and Putnam, H. (1991) 'Unity of Science as a Working Hypothesis', in R. Boyd, P. Gasper, and J. D. Trout (eds), *The Philosophy of Science*, Cambridge, MA: MIT Press, pp. 405–28.

Ordinario, E., Han, H. J., Furuta, S., Heiser, L. M., Jakkula, L. R., Rodier, F. et al. (2012) 'ATM Suppresses SATB1-induced Malignant Progression in Breast Epithelial Cells', *PLoS One*, 7: e51786.

Osborne, J. W. and Overbay, A. (2004) 'The Power of Outliers (and Why Researchers Should Always Check for Them)', *Practical Assessment, Research and Evaluation*, 9: 1–12.

Osimani, B. (2013) 'Until RCT Proven? On the Asymmetry of Evidence Requirements for Risk Assessment', *Journal of Evaluation in Clinical Practice*, 19: 454–62.

Pearl, J. (2000) *Causality. Models, Reasoning, and Inference*, 2nd edn, New York: Cambridge University Press, 2009.

Peirce, C. S. (1892) 'The Doctrine of Necessity Examined', in N. Houser and C. Kloesel (eds), *The Essential Peirce*, vol. I, Bloomington, IN: Indiana University Press, 1992, pp. 298–311.

Peirce, C. S. (1910) 'The Doctrine of Chances, with Later Reflections', in J. Buchler (ed.), *Philosophical Writings of Peirce*, New York: Dover, 1965, pp. 157–73.

Plato, *Republic*, trans. R. Waterfield, Oxford: Oxford University Press, 1993.

Pompa, L. (1967) 'Family Resemblance', *Philosophical Quarterly*, 17: 63–9.

Popper, K. (1959) *The Logic of Scientific Discovery*, New York: Routledge, 2002.

Popper, K. (1962) 'Science: Conjectures and Refutations', in *Conjectures and Refutations: The Growth of Scientific Knowledge*, New York: Basic Books, pp. 43–78.

Popper, K. (1972) 'The Bucket and the Searchlight: Two Theories of Knowledge', in *Objective Knowledge: An Evolutionary Approach*, Oxford: Oxford University Press, pp. 341–61.

Popper, K. (1982) *The Open Universe: An Argument for Indeterminism*, New York: Routledge.

Popper, K. (1990) *A World of Propensities*, Bristol: Thoemes.

Price, H. (1996) *Time's Arrow and Archimedes' Point*, Oxford: Oxford University Press.

Price, H. (2007) 'Causal Perspectivalism', in H. Price and R. Corry (eds), *Causation, Physics, and the Constitution of Reality: Russell's Republic Revisited*, Oxford: Oxford University Press, pp. 250–92.

Price, H. and Corry, R. (2007a) 'A Case for Causal Republicanism', in H. Price and R. Corry (eds), *Causation, Physics, and the Constitution of Reality: Russell's Republic Revisited*, Oxford: Oxford University Press, pp. 1–10.

Price, H. and Corry, R. (eds) (2007b) *Causation, Physics, and the Constitution of Reality: Russell's Republic Revisited*, Oxford: Oxford University Press.

Price, H. and Weslake, B. (2009) 'The Time-Asymmetry of Causation', in H. Beebee, C. Hitchcock, and P. Menzies (eds), *The Oxford Handbook of Causation*, Oxford: Oxford University Press, pp. 414–43.

Price, R. (1763) 'Preface to an Essay Towards Solving a Problem in the Doctrine of Chances', in R. Swinburne (ed.), *Bayes's Theorem*, Oxford: Oxford University Press, 2002, pp. 122–5.

Psillos, S. (2002) *Causation and Explanation*, Chesham: Acumen.

Psillos, S. (2010) 'Causal Pluralism', in R. Vanderbeeken and B. D'Hoodghe (eds), *Worldviews, Science and Us: Studies of Analytic Metaphysics*, London: World Scientific, pp. 131–51.

Psillos, S. (2011) 'The Idea of Mechanism', in P. M. Illari, F. Russo, and J. Williamson (eds), *Causality in the Sciences*, Oxford: Oxford University Press, pp. 771–88.

Putnam, H. (1975) *Mathematics, Matter and Method*, Cambridge: Cambridge University Press.

Quine, W. v. O. (1951) 'Two Dogmas of Empiricism', *Philosophical Review*, 60: 20–43.

Quine, W. v. O. (1960) *Word and Object*, Cambridge, MA: MIT Press.

Quine, W. v. O. (1995) *From Stimulus to Science*, Cambridge, MA: Harvard University Press.

Railton, P. (1978) 'A Deductive-Nomological Model of Probabilistic Explanation', *Philosophy of Science*, 45: 206–26.

Ramsey, F. P. (1928) 'Universals of Law and of Fact', in D. H. Mellor (ed.), *Philosophical Papers*, Cambridge: Cambridge University Press, pp. 140–4.

Ramsey, F. P. (1929) 'General Propositions and Causality', in D. H. Mellor (ed.), *Philosophical Papers*, Cambridge: Cambridge University Press, pp. 145–63.

Reichenbach, H. (1949) *The Theory of Probability*, Berkeley, CA: University of California Press.

Reichenbach, H. (1956) *The Direction of Time*, Berkeley, CA: University of California Press.

Reiss, J. (2012) 'Causation in the Sciences: An Inferentialist Account', *Studies in History and Philosophy of Biological and Biomedical Sciences*, 43: 769–77.

Reiss, J. (2015) *Causation, Evidence, and Inference*, London: Routledge.

Reutlinger, A., Schurtz, G., and Hüttemann, A. (2015) 'Ceteris Paribus Laws', *Stanford Encyclopedia of Philosophy* (Fall), Edward N. Zalta (ed.), <http://plato.stanford.edu/archives/fall2015/entries/ceteris-paribus/>.

Rocca, E. (2017) 'Bridging the Boundaries between Scientists and Clinicians. Mechanistic Hypotheses and Patient Stories in Risk Assessment of Drugs', *Journal of Evaluation in Clinical Practice*, 23: 144–20.

Rocca, E. and Andersen, F. (2017) 'How Biological Background Assumptions Influence Scientific Risk Evaluation of Stacked Genetically Modified Plants: An Analysis of Research Hypotheses and Argumentations', *Life Sciences, Society and Policy*, 13, doi: 10.1186/s40504-017-0057-7.

Rocca, E. and Anjum, R. L. (forthcoming) 'Why Causal Evidencing of Risk Fails: An Example from Oil Contamination', *Ethics, Policy and Environment*.

Rocca, E., Anjum, R. L., and Mumford, S. (forthcoming) 'Causal Insight from Failure', in A. La Caze and B. Osimani (eds), *Uncertainty in Pharmacology: Epistemology, Methods and Decisions*, New York: Springer.

Rosen, J. (2016) 'Research Protocols: A Forest of Hypotheses', *Nature*, 536: 239–41.

Rothwell, P. M. (2005) 'External Validity of Randomised Controlled Trials: "To Whom Do the Results of This Trial Apply?"', *Lancet*, 365: 82–93.

Ruby, J. (1986) 'The Origins of Scientific "Law"', in F. Weinert (ed.), *Laws of Nature: Essays on the Philosophical, Scientific and Historical Dimensions*, Berlin: de Gruyter, 1995, pp. 289–315.

Russell, B. (1913) 'On the Notion of Cause', in *The Collected Papers of Bertrand Russell*, vol. 12, London: Routledge, 1992, pp. 193–210.

Russell, B. (1918) 'The Philosophy of Logical Atomism', in *The Collected Papers of Bertrand Russell*, vol. 8, London: Routledge, 1986, pp. 160–244.

Russell, B. (1948) *Human Knowledge: Its Scope and Limits*, London: George Allen and Unwin.

Russo, F. and Williamson, J. (2007) 'Interpreting Causality in the Health Sciences', *International Studies in the Philosophy of Science*, 21: 157–70.

Saarimaa, H. (1976) 'Combination of Clonidine and Sotalol in Hypertension', *British Medical Journal*, 1: 810.

Sackett, D. L. and Rosenberg, W. M. (1995) 'The Need for Evidence-Based Medicine', *Journal of the Royal Society of Medicine*, 88: 620–4.

Sackett, D. L., Rosenberg, W. M., Gray, J. M., Haynes, R. B., and Richardson, W. S. (1996) 'Evidence-Based Medicine: What It Is and What It Isn't', *British Medical Journal*, 312: 71.

Salmon, W. (1984) *Scientific Explanation and the Causal Structure of the World*, Princeton, NJ: Princeton University Press.

Salmon, W. (1998) *Causality and Explanation*, Oxford: Oxford University Press.

Schaffer, J. (2003) 'Is There a Fundamental Level?', *Noûs*, 37: 498–517.

Schaffer, J. (2004) 'Causes Need Not Be Physically Connected to Their Effects: The Case for Negative Causation', in C. Hitchcock (ed.), *Contemporary Debates in Philosophy of Science*, Oxford: Blackwell, pp. 197–216.

Schaffer, J. (2005) 'Contrastive Causation', *Philosophical Review*, 114: 297–328.

Schaffer, J. (2012) 'Causal Contextualism', in M. Blaauw (ed.), *Contrastivism in Philosophy*, London: Routledge, pp. 35–63.

Shoemaker, S. (1980) 'Causality and Properties', in *Identity, Cause and Mind*, expanded edition, Oxford: Oxford University Press, 2003, pp. 206–33.

Schork, N. J. (2015) 'Personalized Medicine: Time for One-Person Trials', *Nature*, 520: 609–11.

Scott, I. A. and Guyatt, G. H. (2010) 'Cautionary Tales in the Interpretation of Clinical Studies Involving Older Persons', *Archives of Internal Medicine*, 170: 587–95.

Shields, K. E. and Lyerly, A. D. (2013) 'Exclusion of Pregnant Women from Industry-Sponsored Clinical Trials', *Obstetrics and Gynecology*, 122: 1077–81.

Simon, H. A. (1969) *The Sciences of the Artificial*, Cambridge, MA: MIT Press.

Singh Chawla, D. (2016) 'How Many Replication Studies Are Enough?', *Nature*, 531: 11.

Skaftnesmo, T. (2009) *Bevissthet og Hjerne. Et Uløst Problem*, Oslo: Antropos Forlag.

Skolimowski, H. (1966) 'The Structure of Thinking in Technology', *Technology and Culture*, 7: 371–83.

Skyrms, B. (1977) 'Resiliency, Propensities, and Causal Necessity', *Journal of Philosophy*, 74: 704–13.

Smart, J. C. C. (1973) 'An Outline of a System of Utilitarian Ethics', in J. C. C. Smart and B. Williams (eds), *Utilitarianism: For and Against*, Cambridge: Cambridge University Press, pp. 1–74.

Smith, G. C. and Pell, J. P. (2003) 'Parachute Use to Prevent Death and Major Trauma Related to Gravitational Challenge: Systematic Review of Randomised Controlled Trials', *British Medical Journal*, 327: 1459.

Sober (1999) 'Instrumentalism Revisited', *Critica*, 91: 3–39.

Spinoza, B. (1677) *The Ethics*, ed. R. H. M. Elwes, New York: Dover, 1955.

Stalnaker, R. C. (1968) 'A Theory of Conditionals', in N. Rescher (ed.), *Studies in Logical Theory*, New York: Springer, pp. 98–112.

Steurer, J., Schärer, S., Wertli, M., and Miettinen, O. (2014) 'Are Clinically Heterogeneous Studies Synthesized in Systematic Reviews?', *European Journal for Person-Centered Health Care*, 2: 492–6.

Steward, H. (2012) *A Metaphysics for Freedom*, Oxford: Oxford University Press.

Strawson, G. (2008) *Real Materialism and Other Essays*, Oxford: Oxford University Press.

Suárez, M. (2013) 'Propensities and Pragmatism', *Journal of Philosophy*, 110: 61–92.

Suárez, M. (2014) 'A Critique of Empiricist Propensity Theories', *European Journal for Philosophy of Science*, 4: 215–31.

Suárez, M. and San Pedro, I. (2011) 'Causal Markov, Robustness and the Quantum Correlations', in M. Suárez (ed.), *Probabilities, Causes and Propensities in Physics*, Dordrecht: Springer, pp. 173–93.

Suppes, P. (1970) *A Probabilistic Theory of Causality*, Amsterdam: North-Holland.

Suryanarayanan, S. (2013) 'Balancing Control and Complexity in Field Studies of Neonicotinoids and Honey Bee Health', *Insects*, 4: 153–67.

Swinburne, R. (ed.) (2002) *Bayes's Theorem*, Oxford: Oxford University Press.

Tooley, M. (1977) 'The Nature of Laws', *Canadian Journal of Philosophy*, 74: 667–98.

Trefil, J. (2002) *Cassell's Laws of Nature*, London: Cassell.

van Cleve, J. (2001) 'C. D. Broad', in A. P. Martinich and D. Sosa (eds), *A Companion to Analytical Philosophy*, Oxford: Blackwell, pp. 57–67.

van Fraassen, B. C. (1980) *The Scientific Image*, Oxford: Oxford University Press.

van Spall, H. G., Toren, A., Kiss, A., and Fowler, R. A. (2007) 'Eligibility Criteria of Randomized Controlled Trials Published in High-Impact General Medical Journals: A Systematic Sampling Review', *JAMA—Journal of the American Medical Association*, 297: 1233–40.

Venn, J. (1876) *The Logic of Chance*, 2nd edition, London: Macmillan; reprint, New York: Chelsea Publishing, 1962.

Vetter, B. (2015) *Potentiality: From Dispositions to Modality*, Oxford: Oxford University Press.

Vigen, T. (2015) *Spurious Correlations*, New York: Hachette.

von Mises, R. (1957) *Probability, Statistics and Truth*, New York: Macmillan.

von Neumann, J. (1955) *Mathematical Foundations of Quantum Mechanics*, Princeton, NJ: Princeton University Press.

von Wright, G. (1971) *Explanation and Understanding*, London: Routledge and Kegan Paul.

Waitukaitis, S. R. and Jaeger, H. M. (2012) 'Impact-Activated Solidification of Dense Suspensions via Dynamic Jamming Fronts', *Nature*, 487: 205–9.

Webster, G. and Goodwin, B. (1996) *Form and Transformation: Generative and Relational Principles in Biology*, Cambridge: Cambridge University Press.

Whitehead, A. N. (1929) *Process and Reality*, New York: Macmillan Company/Free Press, 1978.

Williams, B. (1972) *Morality: An Introduction to Ethics*, New York: Harper and Row.

Williamson, J. (2006) 'Causal Pluralism versus Epistemic Causality', *Philosophica*, 77: 69–96.

Wilson, E. O. (1975) *Sociobiology. The New Synthesis*, Cambridge, MA: Belknap.

Wilson, J. (2016) 'Metaphysical Emergence: Weak and Strong', in T. Bigaj and C. Wüthrich (eds), *Metaphysics in Contemporary Physics*, Leiden: Brill Rodopi, pp. 347–402.

Wittgenstein, L. (1953) *Philosophical Investigations*, Oxford: Blackwell.

Woodward, J. (2002) 'There Is No Such Thing as a *Ceteris Paribus* Law', *Erkenntnis*, 57: 303–28.

Woodward, J. (2003) *Making Things Happen: A Theory of Causal Explanation*, Oxford: Oxford University Press.

Worrall, J. (2002) '*What* Evidence in Evidence Based Medicine?', *Philosophy of Science*, 69: 316–30.

Worrall, J. (2007) 'Why There's No Cause to Randomize', *British Journal for the Philosophy of Science*, 58: 451–88.

Worrall, J. (2010) 'Do We Need Some Large, Simple Randomized Trials in Medicine?', in *EPSA Philosophical Issues in the Sciences*, Dordrecht: Springer, pp. 289–301.

Wright, L. (1973) 'Functions', *Philosophical Review*, 82: 139–68.

Wright, R. (1994) *The Moral Animal: Why We Are the Way We Are*, San Francisco, CA: Vintage Books.

Zhou, W., Pool, V., Iskander, J. K., English-Bullard, R., Ball, R., Wise, R. P. et al. (2003) 'Surveillance for Safety after Immunization: Vaccine Adverse Event Reporting System (VAERS)—United States, 1991-2001', *MMWR Surveillance Summaries*, 52: 1–24.

Index